Dawn —
Do you have the right cell phone?

Jen

Cell Phone Decoder Ring

Discover how to:

- *Select the right cell phone using a 3-step process*

- *Explore features & services within your comfort zone*

- *Feel good about selecting a wireless solution that's right for you*

**By Jen O'Connell,
Wireless Expert**

Published by Voice of Wireless, Inc.

The Cell Phone Decoder Ring
www.CellPhoneDecoderRing.com

Cover design by Carl Swagerty
Photography by Parish Kohanim
Edited by Robin Quinn
Printed by RJ Communications
Printed in the United States of America

Published by Voice of Wireless, Inc.
For more information on Voice of Wireless, Inc.,
visit our website at www.VoiceOfWireless.com.

All Rights Reserved
Copyright © 2007 by Voice of Wireless, Inc.
The Cell Phone Decoder Ring is a trademark of Voice of Wireless, Inc.

Library of Congress Control Number: 2007921374
ISBN – 10: 0-9792692-0-2
ISBN – 13: 978-0-9792692-0-2

All rights reserved. No part of this book may be reproduced or transmitted in any form by any means, electronic or mechanical, including photocopying, recording or by any information storage and retrieval system, without prior permission in writing from the author.

*To friends who are like family
and family who are like friends...
Thank you to all who have helped with this book!*

Contents

Introduction 7
    How the pieces fit together 12

Part 1: What does *"It's the Network"* really mean?   15
    What language do you speak? (The digital technologies)   16
    I have a need for speed! (The speed of data access) 18

Part 2: Explore Features & Services 27
    Determine your Comfort Zone 27
    Features & Services within your Comfort Zone 29
        Basic 31
        Intermediate 33
        Advanced 46
    Rate Plans 53
    Global Roaming 59

Part 3: Explore Devices 63
What's your favorite band? 65
What are your options?............ 68
    Phones 72
    Smartphones/PDA 75
    Wirelessly enable your laptop 78
Accessories 82

Part 4: What Cell Phone Should I Get? 87
    Ask yourself three simple questions 87
    Use a "wireless worksheet" while shopping 95

Part 5: Living in a Wireless World 97
   Wireless Etiquette (a.k.a. The Rules of the Game) 97
   Fashion Do's and Don'ts 100
   Recycle your old phone... 101

Part 6: Resources 105
   Who to know and where to go for more 105

Glossary 107

About the Author 121

Introduction

Congratulations for taking the first step to gaining a better understanding of the ever-changing and sometimes intimidating World of Wireless. This book was created to educate you about how to select the right cell phone and wireless service. I wrote it in a way that would be easy to understand – in real language – so you can learn the basics, and then apply them to your life. Many of the examples I use are taken from my own personal experiences with my friends and family, people who are probably very similar to your friends and family – and to you too! In addition to the basics, this book will also give you some fun, useless trivia that you can quote at parties or client meetings, making people think you are a savvy early adopter (meaning someone who is the first to embrace new technology) or at least a person who talks like one. If you're not interested in being a technology guru, that's OK too. Still, there's more to your cell phone than just keeping it turned on to get a call at the end of soccer practice. Who knows, you might be surprised to find that it's not as complicated as you might have thought and that this new knowledge proves to be quite useful for your own wireless experience!

One thing is for sure, no individual person knows all there is to know about wireless, but I must admit that I do know much that can be of value to you. I have been in wireless product development for 12+ years, working for such companies as Cingular (AT&T), Verizon Wireless, GTE Wireless and Powertel (T-Mobile). Even with my experience, I still learn something new every day! Thankfully, I have a great role model in my grandfather who never stops learning about telecom either. My grandfather retired from Illinois Bell after 40 years installing phone line infrastructure in and around the great city of Chicago. I grew up listening to his stories about dealing with the unions, how he met my grandmother who was a telephone operator, and later how my grandfather led the team to lay telephone cable which operationalized the first cell site in

North America – enabling Alexander Graham Bell's grandson to make the first commercial cellular call at Soldier's Field, marking the beginning of a new era in communications. The year was 1983 and cellular (later referred to as wireless) was born in North America. Sometimes we joke that my grandfather has passed the baton over to me.

Today I know people at all different levels of knowledge about wireless communications. Below are common examples. Maybe you might recognize some of these folks as people similar to yourself or someone you know. There are many more examples I could present, but these are a few just to get us started. I'll also offer suggestions for how each type of person can reach the next level in their own wireless evolution:

- **My Aunt Sandy – does not have a cell phone.** How anyone in today's world can live without a cell phone is beyond me! She's one of the only people I know without a cell phone, and my Mom is nagging her to get one. I'm in full support of that!
    o **Suggestion:** Get a phone, Aunt Sandy! Any kind will do. You should probably start with a basic device costing under $50. Then look for a rate plan with low usage so you can just get used to having it, like one with 200 minutes (that's a little over 3 hours) a month. You can always change your rate plan at any time if you need more minutes, and upgrade your phone when you want to do more. That way, we can feel better when you're driving home late at night.

- **My Friend's Seven-Year-Old Daughter – does not have a cell phone.** I was amazed while visiting my friends Rick and Amanda for a weekend that a sweet voice from the breakfast table said, "Daddy, can I have a cell phone?" Knowing that I was in the midst of writing this book, I got the "do you believe this?" look from my friend who followed it with the question "Are you hearing this?" When asked, little Maria explained that she needed a phone so she could call her daddy when he goes to work. Never mind the fact that she has recently discovered the power of calling her friends to discuss her competitive toy collection.

- o **Suggestion:** This book is all about finding the right cell phone, however for children under the age of 10 years old it may not be appropriate unless your situation dictates otherwise. If so, there are simplified phones available for children to use which can make it easy to keep in touch with the child who has an active social calendar or visits grandparents for the weekend.

- **My Mom of a Year Ago – always kept the phone turned off.** She has a pink RAZR but only had it turned on when she wanted to make a call (which wasn't very often). Then she would turn it back off again when she was done. She didn't want to drain the battery, she claims.
  - o **Suggestion:** Keep your phone on, anyways. Most phone batteries today can last a week or more in "standby" mode. That way, people can get a hold of you and you might find that you like using your cell phone. Eventually your battery will need to be recharged, but not as frequently as some of the older analog phones you may be thinking of.

- **My Mom of Today – keeps the phone on and is actually starting to use it.** Success! OK, being my Mom, she has had the chance to hear me talk a lot about wireless. So... she actually keeps her phone on now and I can get a hold of her, which is great!
  - o **Suggestion:** My mom still doesn't have voicemail set up. She needs to get that taken care of because people other than just my brother and I are starting to call her on her cell phone

and they'd like to leave a message. Also, since her rate plan allows for free long distance, she can call me in another town without having high long-distance charges on her land-line home phone.

- **My Brother – talks over his minutes.** Like most kids (I call my 20-something-year-old 6'2" brother a kid!), he talks A LOT. So much so that when my brother was younger, he was grounded once for having a $250 cell phone bill in a single month (when it was supposed to be only $40). Oops!
    - o **Suggestion:** Switch "over-talkers" (as I like to call them) to Prepaid (pay in advance), which is a cool alternative to having a traditional post paid (pay after the fact) plan. By paying for your wireless minutes in advance, you can limit your use to avoid unexpectedly high phone bills. Once you have used up all your minutes for that month, you either have to cut back on your talking or ask for more minutes. No more second mortgages, at least not for covering the wireless bills!

- **My Cousin Linda – sends text messages.** Like all 20-somethings, Linda has full knowledge of how to text when she's coordinating where to meet up. She also texts her parents to say hello, so they are getting into it as well.
    - o **Suggestion:** Upgrade her phone to have a Mega-pixel camera with video to take better quality pictures (and home movies) and send them to me! That way, I can see pictures of my new God Child once he (or she) is born! Together, we can experience all the fun things her baby will do even though we live in different cities.

- **My Friend Lyall – uses her phone a lot for business while driving.** Owning a real estate company keeps her on the go. She's constantly on the road traveling from one building to the next and brokering deals with her clients over the phone. Waiting to make a call could make or break a deal... and some places now ask that you keep both hands on the wheel while driving!
    - o **Suggestion:** Stay safe while driving and use hands-free talking with Bluetooth technology. Lyall just bought a new car that has Bluetooth built in. Now she can have in-car speakerphone conversations while keeping her hands on the wheel.

- **My Dad – checks corporate email on his Blackberry.** He likes to "Have a Corporate Day" as he calls it. My dad gets a ton of emails from co-workers all across the country about various things that need to be addressed quickly, and he's good about keeping up. When Dad's out of the office or in a meeting, he needs to know what's going on and stay informed.
    - o **Suggestion:** In addition to checking email, continue to remain connected to the rest of the world by using the web browser on your Blackberry to check traffic reports on the expressway, or surf a news website for the latest in current events. Catching wind of a competitor announcing a new product or just staying in the know is a powerful business advantage.

- **My neighbor Jody – uses his TREO for entertainment.** Most people think that PDA (Personal Data Assistant) devices should only be used for business. Don't get me wrong, Jody does get his work email on his phone... but he also likes to sing his favorite dance tunes and has a few of them stored on his PDA. Jody mentioned one day that he wants to watch movies on his phone since they are starting to become available for download on memory cards.

- o **Suggestion:** Look at a 3G PDA device that would allow for live streaming radio and the ability to download TV/video clips directly from the phone. That way, Jody could access new stuff whenever he wants.

- **My co-worker Craig – uses his phone as a modem to surf the web on his PC**. He loves to test the limits of any device he can get his hands on. As a result, Craig is not afraid of technology and will continue to find new ways to use his phone. He connected his phone using a data cable to his laptop (and once using Bluetooth as well!) so he could surf the web (when a DSL connection was not available) while traveling on vacation with his family.
  - o **Suggestion:** Invest in a PC card that can access the high speed data network, allowing your phone to be free to receive calls. A PC card fits into a slot on the side of your laptop and can access the wireless network if WiFi is not available. Then you are free to surf and talk at the same time.

It's OK if there were a few things I just mentioned that you have never heard of before or don't fully understand. Together, we'll review certain words, what they mean and what's important to take note of – all to give you a better wireless experience. I have included a Glossary at the end of the book for quick reference as you go.

HOW THE PIECES FIT TOGETHER

At the most basic level, wireless – just like people – has DNA which defines itself. In this case, when describing the DNA of wireless, it stands for Devices, Networks and Applications/Services. Having an understanding of these three parts is the first step to unlocking the mystery of selecting the right cell phone and beginning the journey of expanding your wireless knowledge. Your wireless solution will be a combination of these three parts (Device, Network and Application/Services). But before we get into identifying which phone is right for you, it's important to get a little "Wireless 101" training first.

The images shown below will guide you through the three DNA sections we will discuss throughout the book.

Services

One of the many challenges of understanding wireless is that words may often be used interchangeably – for example, "Service" can mean "network coverage" and it can also mean "services & features" or the things you can access on your phone. The reason it refers to both is because a "Service Provider" (otherwise known as a wireless carrier) offers you applications/things to do... which requires... network coverage. I will be as consistent as possible but know that you may hear some words used differently while out there on your own. Sometimes it may be necessary for you to ask a clarifying question if you're unsure.

Part 1
What does "*It's the Network*" really mean?

Funny, but it all really *does* start with the network (not to sound like those familiar TV commercials – but it's actually true!). It's important to ensure that you select the right wireless carrier – one that offers service where you want to use your phone… at home, at work, and while out and about. If you don't have service/coverage, you won't get the call. If you don't get the call, then what's the point? It's best to check around with friends or members of your family who go to the same places you frequent – and see which providers offer better service/coverage.

"The network" is often considered a confusing aspect of wireless to learn because it's associated with plenty of acronyms. Meanwhile, there are two basic parts to the network that you need to understand so you'll be aware of what you will get as part of your overall wireless solution. The two parts you need to be aware of are: (1) "what language do you speak?" – understanding the differences between the digital technologies; and (2) "I have the need for speed!" – understanding how fast your phone can access the internet across various stages of the network.

Let's look more closely at these two network parts.

WHAT LANGUAGE DO YOU SPEAK?
(THE DIGITAL TECHNOLOGIES)

First, it's important to know there are three main types of digital technologies – think of them as different languages… like French, Spanish and Portuguese – that are currently used (spoken) in North America. They are: (1) "GSM" (French), (2) "CDMA" (Spanish) and (3) "iDEN" (Portuguese). OK, in reality, you don't have to know French, Spanish or Portuguese – this is just a comparison to show how they are basically doing the same thing (talking) but in a different ways (languages/technologies).

Wireless carriers are grouped together by their digital technologies:

- GSM   Cingular/AT&T and T-Mobile   use a SIM chip
- CDMA   Verizon Wireless and Sprint   does not use a SIM chip
- iDEN   Nextel   uses a SIM chip

You can easily tell the type of digital technology a phone "speaks" without getting really technical. There are two digital technologies – GSM and iDEN – that use what's called a "SIM chip" in the phone, which acts like the brain of the device. A SIM chip stores your phone number, voicemails and caller ID, and it can be transferred to different devices if you change phones often. Your number goes where your SIM chip (brain of your device) goes. If you're not sure if your phone uses a SIM chip, look under the battery of your phone. If you see this:

... then your device is GSM or iDEN. If you look under the battery of your phone and don't see a SIM chip, then your device is most likely CDMA.

This graphic shows how the nationwide carriers are aligned based on their digital technology:

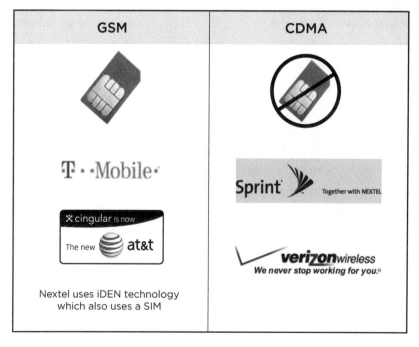

Nextel uses iDEN technology which also uses a SIM

Understanding what type of technology you're using can be of great value if you already have a phone and want to switch to a new service provider. Has this ever happened to you?...

- A new phone on another carrier's store shelf looks just like the one you have in your pocket, but the sales rep tells you, "Sorry – you can't use that phone here."
- A friend gave you their old device because they just got a new one, but it won't work with your current service provider.

Make sure there's a digital technology match between your wireless carrier and the technology of your cell phone. You can't activate a Cingular/AT&T phone on the Verizon Wireless network because they don't speak the same language – even if they look exactly alike. But if you see a phone on eBay that you really want to have, and it's a digital technology match with the wireless carrier that you want to use, that's a good thing!

Note: Before you go full steam ahead with your eBay purchase, there's one little caveat you should know about. GSM devices sometimes have a lock on them for phones used in the US that only lets them work with a specific wireless carrier (also known as "SIM Lock"). If you get a device from a place other than your service provider, be sure the GSM phone is capable of working with the wireless carrier you plan on using. (This rule does not apply to CDMA technology.)

I HAVE A NEED FOR SPEED! (THE SPEED OF DATA ACCESS)

Now we're entering the section which is all about *speed* – how fast you can access data over your phone. If all you ever plan to do is talk, then this section may not be as relevant for you. But if you do plan on learning more of what your phone can do – whether it's "downloading" ringtones or "streaming" last night's *Late Show* on your phone – then this section is important to understand. Later, the things you learn in this section will apply when evaluating devices.

There are some acronyms (yes… more alphabet soup!) that will be explained here and applied again later in this book. Let's get started learning about how the wireless network has a need for speed!

I like to think of the wireless network as a ladder. With each step that you go higher up the ladder, the faster you can access information as you download ringtones, surf the web, or stream video. Each step up on our ladder represents a different speed limit and it will allow you to do more things. Plus, each step builds upon the other like building blocks. Perhaps it might make more sense when you see our ladder. But before we look at it, we need to give each step a name.

You may have heard the terms "2G" or "3G." People talk about this like they're really important, but not everyone understands why they are relevant. Let me try to help you get this part. "G" stands for generation. And it's a reference to the fact that the wireless industry is constantly evolving. "1G" was the beginning. It relates to Analog and is the lowest step on our ladder.

Remember the days of the huge phones that were the size of a brick? (The need to refer to old phones as bricks is because they are considered large and heavy by today's standards.) I remember the old phone my Dad bought me for college. That device used was 1G (using the analog network). Even so… I thought I was the coolest kid in school! Other memories are that of the late 80s and early 90s when phones required those curly cue antennas on the back of cars. After much debate about cross-talk interference, the privacy threat of eavesdropping, and a type of identity theft called "cloning," some really smart people decided it was time to improve upon what we had. We needed to move away from analog (1G) and enter the digital age (2G).

This was also a time when new players started hitting the marketplace and wireless competition began to heat up. The time was 1996 and North America was introduced to "Digital." This is when GSM and CDMA were born in North America. (This phase also included a digital technology called TDMA which is no longer actively supported by today's wireless networks.) These technologies represented the next generation of cellular, so naturally they became 2G. Still, we wanted to make our cell phones become more than just something you could talk on – we wanted a new way to communicate. Soon we discovered text messaging and the internet on our phones. Right then and there, our need for speed began.

The cell phone industry entered a race to be better, quicker, faster. They had their eye on advancing into the world of multimedia (3G). Some technologies were able to get there right away, while others needed a half step to get us on our way. So... a middle step between 2G and 3G was created called 2.5G. Let's see what all of this looks like on our ladder so far.

Network

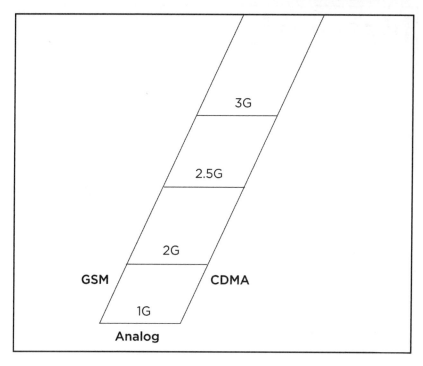

Although we recognize three main digital technologies (GSM, CDMA and iDEN), we will only use the top two for simplicity. Notice how GSM and CDMA are represented on either side of the ladder? Although the two main digital technologies (GSM and CDMA) have a need for speed, they got there in different ways because they speak different languages. By showing our ladder in this way, you can compare digital technologies – looking at the ladder as a side-by-side comparison. Again, this will help you down the road to understand when comparing different devices to see which one is a better fit for you. Let's see what our ladder looks like when we add the names of the 2G, 2.5G and 3G digital technologies as they begin to grow up the ladder to get faster and faster with data speeds.

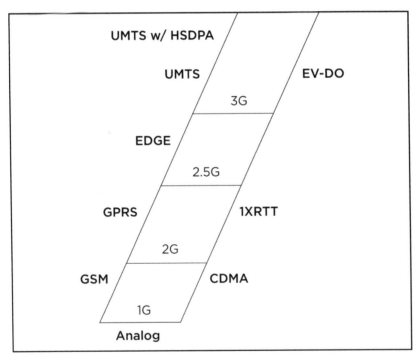

What you see here is a listing of network acronyms that correspond to a technology – adding to our ladder as it evolves for your understanding. GSM has its group on the left and CDMA has its own on the right. The reason for showing this stage of the ladder is to get you comfortable with seeing these terms. I want you to know which acronyms are at similar steps as you travel up the ladder.

As I've mentioned, the higher you go up the ladder, the faster the speed you'll have when surfing the web. This next picture represents the *speed* at which each acronym on our ladder, based on the two digital technologies, can travel. These are ranges and may be different by wireless carrier, but it can give you an average. Keep in mind that the speeds related to this next version of our ladder are similar to the speeds you use while accessing the internet in your own home. 2G = dial-up modems. 2.5G and 3G = DSL lines and cable modems. When we get into 4G and beyond, we'll see even faster speeds by using WiFi.

Let's see what we can expect today.

Network

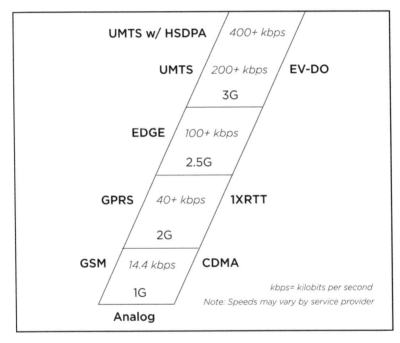

On this ladder, you see the term "kbps"; this stands for kilobits per second. It refers to the amount of data you can send over the internet within one second. The higher up the ladder, the more data you will need to send at one time, which is needed for a good experience when listening to streaming radio or watching TV on your phone (an example of 3G). Lower on the ladder is good enough for downloading a ringtone or surfing the web (an example of 2G). We will get more into this when we review applications later, but it's important to understand these terms now. They are used quite often in TV commercials when comparing wireless services and on phone boxes in the technical specifications section. This knowledge is important when selecting a rate plan or determining what you want to do on your phone. You don't need to know what the acronym "HSDPA" actually stands for (by the way, it stands for High Speed Data Packet Access), but if you know you want to stream video, and the device you have only supports GPRS, then you may want to find another device higher up the ladder to ensure you have enough speed to accomplish what you want to do. On the flip side, if you're looking for something new and the sales rep wants to sell you the newest EV-DO device and all you want to do is make phone calls – then maybe you're getting yourself into something more powerful than what you need.

Whew! The hard part is over... now let's talk about the fun stuff!

Part 2

Explore Features & Services

We've covered networks, so now let's talk about what you want to do on your phone.

As you can see, the level of Network functionality (as shown on our ladder) determines which service is offered (for the best experience). Lower on the ladder, the more basic the service. The higher up the ladder, the more advanced the service. Thus groups called Basic, Intermediate and Advanced are formed. The best way to learn what's available on your phone is to find the group that you feel comfortable in and then explore what's within your comfort zone. Chances are that you never realized all the different services that exist on your phone and within your comfort zone.

DETERMINE YOUR COMFORT ZONE

By matching yourself up to a group, you can learn of different features & services that may be within your comfort zone. These features & services can allow you to maximize your wireless experience. Also remember that not every device can do everything available today, so also be mindful of your existing device's capabilities as you dream of what's next for you. Let's explore each of the three Comfort Zone groups in more detail...

Description of Comfort Zone Groups

Take a minute to determine which group sounds more like how you use your phone today or how you would like to use your phone going forward.

COMFORT ZONE	DESCRIBE THE RELATIONSHIP YOU HAVE WITH YOUR PHONE	HOW YOU CAN USE YOUR PHONE
Advanced	"I create an experience on my phone." (Entertain me!)	You bring content to your phone by watching TV or listening to music. It's very entertaining!
Intermediate	"I use what's on my phone." (I can communicate without talking.)	You use some of the features & services available, such as text messaging or taking pictures using the camera. Email access may also be important.
Basic	"I've got a phone." (It's all talk.)	You use your phone to make and receive phone calls, that's about it.

Once you've selected your Comfort Zone, let's apply that knowledge back to our network ladder to see how the two relate to each other. This will help us further still when we move to the next section for selecting a device. Whether you are aware of it or not, you have already narrowed down the playing field since you should focus on selecting a device that is compatible with your Comfort Zone. You will need to take note of the network technologies (CDMA or GSM acronyms) that are at the same level as your Comfort Zone (Basic, Intermediate and Advanced). See why we didn't jump right into the device section first?

Let's take a look at where you are at so far:

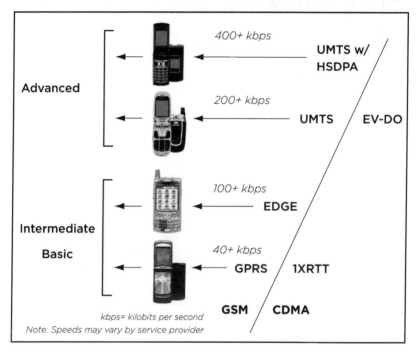

FEATURES & SERVICES WITHIN YOUR COMFORT ZONE

This next section explores the various *features* (things on your phone that don't require an extra fee to use) and *services* (things you pay extra for) that you can access on your device. Before selecting the right phone, it's important for you to know just how you want to use it. When it becomes time to select a device, you will want to be sure it supports all the features & services you're looking for. Here's an example of a few features & services – by Comfort Zone group – to get you started. There are many more for each group. They may be easier to find now that you know what you are looking for:

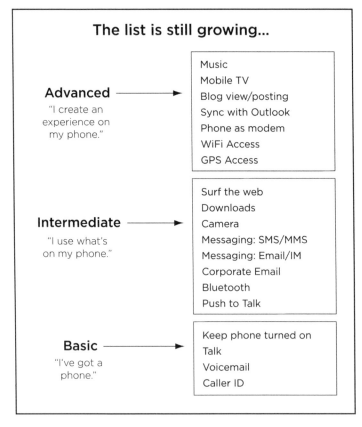

Note: If you're already in the Advanced group and have used all the features & services I list in this book, rest assured that every day something new is being developed to continually push the envelope of what's possible.

With that being said, let's review these features & services – by Comfort Zone group – for a better understanding of each.

BASIC

Everyone starts out as a basic user whether they want to admit it or not. Just like how we noticed the network ladder has building blocks, think of feature & services usage in the same way – starting with the Basic category as our foundation.

Keep phone turned ON

Many people use their cell phone only in emergencies. Within this group, many of the individuals keep their cell phones mostly turned off. Well, what about those folks who may need to contact you? When your phone is turned off, it doesn't allow others to get a hold of you. I suggest that you always keep your phone on. If you don't want to be bothered, then put them into voicemail.

Voicemail

Setting up your voicemail is one of the first things you should do when you get wireless service. It's basically expected that you have this available so when others call you and you're not around they can leave you a message. That way, you can call the person back when you're ready to talk, which is also considered good cell phone etiquette.

SERVICE: You will be charged minutes to access your voicemail with your cell phone. Or you can call your voicemail from a landline phone if you wish to avoid cell phone charges.

Talk

The Basic Comfort Zone group centers on using the phone to make and receive phone calls. It's the start of creating a lifeline and a way to stay connected wherever you go. When you talk on the cell phone, you speak with your voice, so often the term "talking" may also be referred to as "voice" when it comes to describing how you want to use your phone as well as the rate plan associated with that action.

SERVICE: You will be charged per the minutes you use. Rate plans offer many choices to appeal to those who want to talk a little or a lot. I'll cover various rate plan options at the end of this section.

Caller ID

You decide which numbers are worth remembering, by programming them into your phone's address book. (Since Caller ID does not exist the way it does on your home phone, just a number will show up until you decide to give it a name.) You can either save a number that has called you or enter a number from scratch by assigning any name you choose to give it. Since many people have multiple numbers, it may make sense to add a descriptor matched to home, work and mobile with the name. Other times, nicknames or assigning pictures or unique ringtones to Caller ID can make your experience fun when someone calls you.

FEATURE: Caller ID does not require an additional fee to use.

INTERMEDIATE

The next Comfort Zone group comes in the form communicating in ways other than with your voice at the Intermediate level by taking advantage of the applications that already exist on most newer phones from the past few years. It enriches your mobile experience by adding a level of convenience... at least that's the way people who use data view it. If you've never used data on your phone, then it may look more like a bother. But try it before you knock it. Chances are that you access these services already – so wireless data is nothing more than accessing what you already use on your laptop, on your cell phone too.

Surf the web

I'm not exactly sure how the term "surf the web" became a metaphor for accessing information over the internet, but just about everyone seems to know what it means. There are two ways to access the internet on your phone. One way is to access a page called a "portal" that gives you suggestions of places to go based on what you want to do. Often, you can personalize this page so you get instant access to your news, weather, sports, etc. The other way is when you know exactly where you want to go and you type in the web address or URL. Not all websites are optimized to show up on a cell phone, but many are. Get familiar with the options and menus while in your phone's web browser to see what's available. To access the web browser, either check the main menu in your phone or often you will see an icon on your phone keypad that will launch the browser for you.

It's also important to know that having access to the web on your phone enables far more than just access to content and web pages. The web is also the way to access email, instant messaging, downloads, streaming video and listening to streaming radio. It can open up your phone to greater possibilities once you're ready for them. (We'll review more of what you can do from the web as we continue through the Comfort Zone groups.)

SERVICE: If you plan on using the web, then it will be in your best interest to look at getting a data rate plan that supports the appropriate level of usage. Talk to your sales rep about which plan is right for you or refer to the rate plan section later in the book.

Downloads

This is a way to shop for new stuff from the web to put on your phone. Downloads consist of a variety of options from ringtones, games, pictures and even music. The list can go on and on.

Another trend is having "ringtones" that a person hears when they call you, which are known as Answer Tones or Ringback Tones (depending on your wireless carrier). Instead of hearing the tone known as "It's ringing and I'm waiting for you to answer," the caller listens to a nice song until you are able to answer the call. I have one on my phone – it's a Jessica Simpson song and everyone comments that it's a snappy tune to listen to until I answer. You can also personalize the song to an individual caller – my brother hears the theme to *Superman* when he calls me.

SERVICE: You will need access to the web on your phone to search for the right download. Each item you wish to download will have either a one-time fee or a monthly subscription associated with it, depending on the type of download. If you don't have access to the web on your phone, you can also purchase a download from your laptop and send it to your phone within a message.

Camera

Most phones now have cameras in them so you can take a photo spontaneously. People started to realize the value of having such a camera when they noticed that their cell phone was always with them, yet somehow they left their digital camera at home – or simply didn't want to carry an extra device with them to an event. The level of sophistication is getting better as technology improves. The lesser quality option is "VGA." The more sophisticated and better quality alternative is the "mega pixel" (MP). With mega pixel cameras, the image will be sharper when there are more pixels your camera can capture. Therefore, the higher the number mega pixel, the better your image will continue to be (as is true with regular digital cameras).

One thing to note, as the camera quality increases to 3 mega pixel and beyond, for example – be sure that it also has "image stabilization" or the image may look fuzzy due to your hand shaking and the photo will not look that great. If the phone doesn't have this feature on a multi mega pixel device, then look for something lower such as a 1 mega pixel or even a VGA camera. Believe it or not, those images still look great on your phone screen – the only time you will notice a difference in picture quality is when you choose to print the photo. Be sure to try out this feature and compare to make sure you like what you see.

Many businesses are strict about not allowing cell phones with cameras, so you may find devices that support options with or without a camera.

FEATURE: It doesn't cost anything extra to take pictures on your phone. If you have a data cable or Bluetooth, you can transfer your photos to your laptop at no extra cost as well. You can send your photo either in a multimedia message or in an email.

Messaging: SMS/MMS

Those of you who became proficient at the road trip game of "decipher the license plate" will have no trouble with this section. The mid-1990s brought about a new service called "Text Messaging" or "SMS" (Short Message Service) where you can send a little note (comprised of characters – including letters, numbers, spaces and punctuation) from one cell phone to another. "MMS" (Multimedia Messaging Service) is nothing more than adding a picture, video or sound clip to a text message.

The way you would type a message from your phone is similar to how you would dial 1-800-PIZZA-2GO. You would actually dial the letters that correspond to the number keypad. But with messaging, you have to spell out exactly what you are trying to say. The number 2 has the letters A, B and C associated with it. If you want to spell the word CAB for example, you would do the following:
- Press the "2" key 3 times to get the "C"
- Press the "2" key 1 time to get the "A"
- Press the "2" key 2 times to get the "B"

This process of pressing a common button multiple times in a row to get to a certain letter, number or symbol is referred to as "triple tapping." Yes, there is a name for everything!

To send a text message, find Messaging within the main menu of your phone. This section will allow you to do a number of activities, including create a new message. Using the triple tap method described above, you can create a quick note to another cell phone or email address. Upon opening a new message to create your note, it will be similar to opening a new Word document on your computer. It will be a clean slate with a blinking curser. Start typing. If you need to spell something with letters that share the same number key, you will need to pause a second to let the curser advance one space, recognizing that you're ready to move on to the next letter, spelling out what you wish to say.

On the top of most screens while creating a new message, you'll notice a number that goes down as you type. This is a counter to let you know how many characters you can type in a single message. Not all messaging services are created equal, but most support up to 160 characters – some may have more. Even if your phone says it can support more, it may divide your message into more than one message (usually in chunks of 160 characters per message) so what may appear to be a long message to you will actually be received on your friend's device as two or more messages. This is called "truncating."

Because messages are limited by the number of characters and people usually send them while on the go, messages have a tendency to be abbreviated and short. A sort of new language was developed, and it's used in all sorts of quick messaging services, including Instant Messaging and email. See if you can decode some of the fun phrases used:

CU LTR	*"See you later."*
W8 4 Me @ home	*"Wait for me at home."*
LOL	*"Laugh out loud"*
K	*"OK"*

<u>If you only learn one thing from this book – it should be this!</u>

Everyone has a story for this day – here's mine. On a weekend in mid-September 2001, my brother came out to New York to visit me. We hung out in Manhattan, toured our favorite stops in Times Square, and went to a concert at Madison Square Garden. The trip came to a close and it was time for him to fly back home. It was the morning of September 11th, and he had a 9 am United Airlines flight out of Newark Airport. I dropped him off with time to spare. As

I was leaving the airport, my dad called me on my cell phone to be the first to deliver the news. While leaving the Newark airport, I then saw what he was talking about – The World Trade Center tower was on fire! Beyond the news that my dad had reported to me, I then saw a second plane that looked like it was heading toward the second tower. "Huh – my depth perception must be way off this morning," I thought, as I continued driving away from the city and towards my office at the Verizon Wireless Headquarters in New Jersey. You know the rest of the story of what happened that September 11th day, and it had a very profound impact on the world.

Soon there were reports of planes being hijacked, including a United Airlines flight from Newark that took off that morning. Thoughts of my brother, who was dropped off only an hour earlier, overwhelmed me. If only I could call him to see if he was safe – and not aboard that plane!!! Home phone lines as well as cell phones were all jammed with too many people trying to call all at the same time. "Fast busy" or all circuits are busy was a constant annoyance in my ear as I frantically redialed hoping to hear his voice. I could not get a call to go through! If only he knew to send me a text message!

In times when the phone lines are jammed and a call can't go through, a text message can still be delivered. Text messaging uses a different part of the wireless network that does not get congested by an overwhelming volume of calls trying to go through all at once. Messaging uses a part of the network called the "control channel" and it has plenty of room to work – even during times of national emergencies, such as September 11th and Hurricane Katrina, or just at rush hour when you're driving home from the office in traffic. (Note: In order to send a text message, the wireless network still needs to be operational, so make sure you can see good signal strength in order to be certain that your message has been sent.)

By the time I got to the office, text messages were starting to come in for other people. My co-worker Alex got a text message from his wife, who was in the first tower when it was struck. She was able to send a message saying that she was OK after walking down 40 flights of stairs to safety! My brother was quickly educated on this feature once he was able to get a call through saying his flight never took off and to come pick him up. Actually, the events of that day were my first inspiration to write this book! Should you find yourself in an emergency situation in which you cannot make a call, please use text messaging to communicate your location to others as well as your status (that you are safe, need help, whatever). Please inform your friends, family and kids to learn how to do the same.

SERVICE: You will be charged by the number of messages sent and received. If you plan on using this service, then you may want to look into bundles of messages which bring the cost per message down to make it affordable.

Messaging: Email/IM

This "killer app" as it has been called for several years is the popular ability to send/receive personal email on your wireless device – whether that may be your phone or PDA. More and more people are looking to email as a way to communicate. We've become addicted to it, and email often replaces the activity of actually calling someone. Even my Grandmother in Montana uses email. It's a great way to communicate with folks in different time zones, a way to let people know you're thinking of them while you're on a call or a way to keep in touch while you're busy throughout the day. To continue that line of communication from your cell phone is a natural extension of staying in touch. Have you have ever felt cut off from the world when your email goes down? Then you know what I'm talking about.

Like all things wireless, there needs to be compatibility in order to make this solution work.

- Compatibility #1. When thinking about getting email on your phone, you need to first see if your email account can be accessed from places other than from the laptop website. (If the answer is yes, then your email is "POP3" enabled.) Post Office Protocol version 3 (POP3) means that you can access internet email from an alternate source; in this case, your phone would qualify. With some email accounts, this feature is free, and others may charge to become POP3 enabled. Still others are considered proprietary and can only be accessed from a device that is approved to retrieve your internet email. You need to keep this in mind when evaluating which email accounts and internet service providers (ISP) you should use, if you're thinking of signing up for a new email address.

- Compatibility #2. The second element of compatibility is that your phone needs to support the functionality of accessing email. Most of the newer devices can support POP3; others can, but you need to download the email application. Some of the lower end and older devices will not be able to support the use of email. If you see POP3 as a device technical specification on your phone box or sales sheet, then you are good to go.

Another personal email term is called "IMAP4." This feature basically allows folders that you have on your internet email to also be available for viewing and access from your phone.

Another service similar to email but used more for rapid back and forth conversations without keeping an archive of the written word is "Instant

Messaging" (IM). This service is gaining in popularity and at times is being preferred over email for the younger crowd. It takes communicating to the next level by adding an element of availability before you communicate. You can let friends in your Buddy List know when you're free to chat or away. As a frequent user of IM, when you're going to be away from your laptop but still want to be available to your friends, you can take IM on the road with you and access it from your phone. You'll show up in your buddy list as being mobile. Again, your phone needs to support this feature. The more popular IM service providers such as AOL, Yahoo! and MSN are likely to be supported.

SERVICE: Both Email and IM require internet access. If you already have a data plan that supports web browsing, then email and IM are covered under that plan.

Corporate Email

Access to your work email may be a little more involved than your personal POP3 email. That is because most corporations require a layer of security. If you use a "VPN" (virtual private network) then chances are that you'll need to explore secure corporate email options. There are a variety of ways to access secured email but the most commonly known is "Blackberry." Blackberry devices support blackberry service, but other PDAs and devices may also have access to other corporate email applications. This will typically require the help of your company IT professional to ensure that you're set up properly.

Note: Blackberry also provides access to personal POP3 email so you have the ability to use these devices for personal access as well.

SERVICE: Corporate email access is typically a different rate plan than just having standard internet access. Check with your service provider to see what your corporate email access options are. You may be required to download an additional corporate email application if your device does not support corporate email access at the time of purchase.

Bluetooth

One day, my friend Lyall called me saying that she was test-driving a new car and that the salesman – recognizing that my friend is a savvy shopper and does her research before engaging into negotiations – was informing her of all the advanced gadgets and features of her soon-to-be new vehicle. When the sales guy mentioned to my friend that the car had "Bluetooth," he was expecting more than just a blank stare on her face. Immediately, Lyall recognized that this was in fact more than just a conversation about oral hygiene – "can teeth really turn blue?" she thought to herself. "Why in fact would I want that... in a new car?" she asked me.

After having a chuckle at her expense, I explained that Bluetooth is actually a technology that works with your cell phone as well as other devices. When in the car, Bluetooth allows you to keep both hands on the wheel while talking on your cell phone – like having a built-in speakerphone. It uses speakers in your car, so you can hear the person on the other end. When combined with a built-in microphone, that same person can now hear you, completing the speakerphone system without you having to physically touch your cell phone – which will probably still be in your purse or pocket.

Basically, Bluetooth is a technology that gets rid of wires and makes a connection between your cell phone and another device/accessory, for example. When out of the car, you can talk and hear the person by using a Bluetooth headset (wireless headset for your wireless phone) – while having a conversation without anything physically being connected to your phone. Have you seen people walking around looking like they are extras in a *Star Trek* movie?... and talking into what appears to be thin air? That strange thing placed in their ear is most likely a Bluetooth headset, and the conversation is probably nothing more exciting than placing an order for Chinese takeout. Beam me up some sweet and sour chicken, Scottie!

What else can Bluetooth do for you?

You can use Bluetooth to transfer data from one laptop to another device, such as a PDA. It's commonly referred to as "file sharing." An example of this would be placing a PowerPoint presentation on your laptop and then you visit your client's office to give the presentation. Let's say it goes really well, and they would like to get a copy of your PowerPoint file so they can refer back to it at a later time. You can send it to them using Bluetooth, assuming you both have this technology available on your computers. Essentially, the machines will talk to each other and will send a copy from your laptop to your client's laptop. You may be familiar with a similar ability called "IrDA" (Infrared data). That is the small, shiny black rectangle found on laptops, PDAs, remote controls, etc. The same principle applies. Getting really geeky: The difference between IrDA and Bluetooth is that IrDA requires direct line of sight – meaning the two devices need to be looking eye to eye at each other. In contrast, Bluetooth does not require the two devices to look right at each other; just merely that they be within close range (a maximum of 30 feet; some phones may allow you to be ever further away). Also, the presentation we referred to will transfer much faster over Bluetooth than over IrDA. This is the equivalent of driving 65 mph on the highway using Bluetooth vs. 35 mph on local streets with IrDA.

Printers are now enabling Bluetooth. You can send pictures taken from your Bluetooth capable camera phone and send them to a Bluetooth capable color printer, so you can have something to put in a frame. Cool, huh?

Like any good technology, Bluetooth is constantly advancing in its capabilities. The next big thing is the ability to listen to music over a Bluetooth stereo headset. So you can *Star Trek* yourself out while jamming to your favorite tunes, assuming that you have a Bluetooth-compatible MP3 player. (No worrying about getting tangled up in your headphone wires while jogging through Central Park.) This Bluetooth profile is referred to as "A2DP" (no, it's not a friend of R2D2 or C-3PO's cousin), and it's finding its way into newer Bluetooth phones that have music capabilities.

I now pronounce you – cell phone and headset – you may pair

Going back to the basics again… there is one crucial step that needs to be performed before you can get your Bluetooth capable device and headset to play nice together in the same sandbox. This is done by creating an exclusive relationship, like a marriage, between your Bluetooth capable phone and your Bluetooth headset. (Note: If you have a fear of commitment, you may want to stick to the old fashion wired ear bud.)

The marriage between the two is necessary so when your phone rings, it can only be answered by YOU – not the guy on the bench sitting next to you, who also has a Bluetooth headset. The task of making sure your phone and headset (or any two devices talking via Bluetooth for that matter) work only with each other is called "pairing." The way to establish this pairing which can be cancelled at any time and re-established as many times as you like… (without the need for a divorce lawyer) may vary depending on the devices you're working with, so always refer to your user's manual if the steps are not intuitive to you at first. On a cell phone, you can usually get set up by going into your connections or settings menu. Once you find the Bluetooth section (if not found, check to make sure that your device supports this technology), you'll need to turn the feature on. Next, you'll need to get your phone to look for all the Bluetooth devices that are close by. (Please – don't attempt to do this in an airport as you may find more than what you bargained for! It's best done in your home or car.) The name/model number of your headset, for example, will show up as being found by your phone. To personalize the name of your headset so it's easy to find while pairing, you can give it a name by typing in something such as "Jenz Headset"... from your phone. It will assign the name to your headset until you decide to change it. A password or PIN will be required to complete the pairing. The default is always: "0000". Once you have made your claim to the headset, you can change the password or PIN so only you can have access to it.

How Bluetooth got its name

At this point, you may be asking yourself, "What's up with that crazy name of Bluetooth anyway?" The story is told that the guys who developed this technology of "wirelessly transferring voice or data from one device to another" realized that they were really on to something here. They were going to make a ton of cash licensing out the rights to use it, and were going to bring riches back to their homeland of Scandinavia. A similar idea to what the Viking Pirate had set out to do centuries earlier. His name was Pirate Bluetooth. *Get it?*

FEATURE: It's important to know that you're not charged for using Bluetooth technology itself, but your minutes still apply when talking over a Bluetooth headset. If you use Bluetooth to connect to your laptop for sharing a picture or file, then there's no charge, since you're not accessing the wireless network. Instead, you are sending data from machine to machine just like IrDA.

Push to Talk (PTT)

"Push to Talk" (PTT as it's referred to commonly) got its start with Nextel. Soon after it proved its success, everyone wanted to try it. This service allows you to convert your phone into a walkie-talkie that helps in coordinating with large groups of people all at once or via speakerphone, allowing you to talk to a friend – often using a different bucket of minutes than what would be used when making a traditional cell phone call, depending on your service provider. Other service providers have "availability" built in so you can see of your friend or co-worker is available to speak (just like in your Instant Message buddy list) before making a PTT call.

SERVICE: Calls made over Push to Talk are charged differently than making a regular cell phone call. You'll need to check with your service provider to see the plans which are available.

ADVANCED

Music

Phones that play music have been around for a few years. They're commonly referred to as having an MP3 player in the technical specifications. For those of you who don't know, "MP3" is nothing more than a audio file format that can be played by a computer or another device. Similar to how you can read Word documents or Excel, MP3 is a software file that's dedicated to sound for playing music, instead of showing words or pictures. MP3's are usually downloaded from the internet on your computer or they can be ripped (copied) from an existing CD. Other music file formats exist, such as "AAC" (which is the file format that Apple's iPod can play). Whether it's MP3, AAC or other formats, having the ability to take your music with you while on the go brings new meaning to the term "converged" devices – meaning merging the functionality of different devices into one.

Music has its own set of rules and functionality. Let's review:

- Sideload – This is the term given to the task of copying/moving files from the music library you have on your laptop and then transferring those same files to your phone. The transfer is typically done by use of a data cable, and then the music is stored on a memory card in your phone. Since most songs are about 4 MB, you need to be mindful of how many you can transfer at one time so you don't exceed the available memory. (For example: If you have a 64 MB memory card, you can expect to store roughly 16 songs or a full CD.)

- DRM – "Digital Rights Management" is a way for Record Companies to monitor music and other content by limiting the distribution of that file. This is mostly in place to ensure artists are properly compensated for their songs, including ringtones as well as transferred music. Depending on where you obtain your music, there may be certain restrictions on how you can transfer it and DRM helps to regulate this. You may want to look for a device that supports "Windows Media Digital Rights Management" (WMDRM) for the most open form of transferring protected content, which can support Napster-like content.

- OTA – "Over-the-Air" is a means to transfer a file from one place directly to your phone which may or my not involve the web. In this case, we are referring to the purchase of a song over the web so it can be placed directly into the music library on your phone. It's best to have a 3G device so it doesn't take as long for the music files to download.

- <u>Streaming Radio</u> – When accessing the web, you may be able to listen to live internet radio from your phone. One thing to note is that when you're doing so, your phone calls may go to voicemail – but at least you'll have access to a limitless list of songs. Check to see if your service provider will allow you to still get caller ID if a call comes through while you're listening to internet radio, so you don't miss a call.

Note: Depending on what you're doing with music, it may not cost anything to use. Below shows both examples:

FEATURE: If you transfer music from your laptop to your phone (using a data cable or Bluetooth) and listen from the phone's music player – no additional rate plans are needed.

SERVICE: If you plan on downloading songs from the phone's browser (additional fees per song may apply) or listening to live radio over the web, then you'll want access to the wireless internet from your phone. I would strongly encourage an unlimited data rate plan.

Mobile TV

There's a term that is being used and it's called the "4th screen." It is in reference to the way people are entertained. The 1st screen is the cinema or movies. The 2nd screen is TV. The 3rd screen is your laptop/PC. And the 4th screen is your cell phone, where you can access video or TV. There are two different ways to obtain programming on your phone, assuming your phone supports this feature. (Not to be confused with the other example of "3 Screens" – this is viewing video/TV content on your TV, laptop and phone for combined service delivery by companies such as AT&T and Verizon. We'll touch on this later.)

1. <u>Downloading program clips</u> – Previously recorded programs and movie trailers, etc. can be available for download to your phone to watch when it's convenient for you. You may experience "buffering" (when your phone will start and stop playback as it's trying to receive the show and play as much as it can before it needs to go back to the internet to ask for more). For this, it's best to have a 3G device so you can download at a faster speed. Also, you'll need to be mindful of the storage capacity on your device to ensure that you do not exceed your available limit so the clip has a place to reside on your phone.

2. <u>Stream live broadcasts</u> – Some news, weather and other programming allow for their broadcasts to be available as live TV on your phone, similar to watching certain TV shows at home. This is a convenient way to stay in the loop or entertained while on the go.

SERVICE: In order to get access to various TV programs from your phone, you'll need access to the wireless internet (I strongly encourage an unlimited plan) and a compatible device. Check with your service provider to see if there are any premium fees for the channel you wish to access.

<u>Blog view/Posting</u>

If you're part of the MySpace generation or like to stay connected with your friends and family who live far away, posting pictures taken from your phone directly to your blog site is the coolest in wireless convenience. You can also log onto your blog site to check comments or post a new comment for someone else. Welcome to the social networking wireless generation!

SERVICE: In order to access a blog site or post to it, you'll need a data rate plan with access to the internet. If you already have access, then accessing blogs should be covered.

Sync with Outlook

If you use Microsoft Outlook on your laptop and want to access your calendar while away from the PC, then you can "sync" your calendar, contacts, address book and tasks to your phone. (Note: Check to ensure that your device supports this feature.) Many of you may know this feature is available on PDA devices, but you might be surprised to know that many standard phones support this feature as well. It may require additional software so your laptop knows how to send the info to your phone. This software is often available for sale or download on handset manufacturer's websites.

FEATURE: If you sync your content over a data cable or Bluetooth, there are no charges. Your information will only be as recent as the last time you sync'd.

SERVICE: If you sync your content over the air (for real time updates) then you'll most likely need a data rate plan with access to the internet. Check with your service provider to see what options are available for real time sync of your outlook content.

Phone as a Modem

You may remember an example from the beginning of this book of a co-worker of mine who used his phone to provide access to the internet so he could surf the web and check email from his laptop while on vacation with his family. Most phones are capable of this, but you'll need a data cable (or Bluetooth) and software (you will need to download "drivers") to make sure your phone and laptop know how to talk to each other. You can get access to this software from most handset manufacturer's website or from your wireless carrier. It's also best to ensure that your device is higher up the network ladder since that will dictate your internet speed. It's a great option to stay connected while traveling or "working" from the beach! (Yeah – right!)

SERVICE: You'll need access to a data rate plan. Some service providers may require you to have an additional rate plan specifically for a laptop connection. Check with your service provider. If you already have a data cable to sync your content, you can use that same data cable to connect your phone to your laptop for use as a modem, as well.

WiFi Access

In case you were wondering about WiFi, it's the kind of thing you hear a lot about in Starbucks, airports and other places. It's the same technology that people use in their homes when they connect their cable modem to a little box and... presto! They can walk around the house with their laptop still connected to the internet. WiFi is different from the cellular network but can be found on some cell phones. It's also being considered as the future "4G" on our network ladder – so you'll be hearing a lot more from WiFi (or WiMax) in times to come!

In the meanwhile, there's an entire alphabet soup for this technology (different from CDMA and GSM), so I'll list a breakout below so you can understand what these letters and numbers mean:

WiFi	SPEED	FREQUENCY
802.11b	11 mbps (11,000 kbps)	2.4 GHz (2,400 MHz)
802.11g	54 mbps (5,400 kbps)	2.4 GHz (2,400 MHz)
802.11a	54 mbps (5,400 kbps)	5 GHz (5,000 MHz)

Instead of using acronyms to climb the network ladder, we have number and letter combinations instead. The same rule for speed, as we found on the network ladder also applies here as well. Soon you will learn about frequency for phones, and WiFi uses this as well. Cellular and WiFi have a lot of similarities and they have the same common elements, just with different names, that access faster speeds at different frequencies. (You may want to come back to this section once you've read through the device section.)

SERVICE: There are two options for WiFi access: (1) You can subscribe to a public WiFi "hotspot" network directly from your laptop in locations such as Starbucks and the airport. If you have WiFi on your PDA device, then check with your service provider to see if a public WiFi rate plan is available; and (2) The private WiFi access you may have in your home is yet a different option. You may be able to detect both options on your PDA or laptop. However, when it comes to access, you may have to select which network path is appropriate for you.

GPS Access

"Global Positioning Service" (GPS) is a technology that allows your phone to talk to a satellite to find out exactly where you're located on the globe. GPS can have many uses, including tracking kids after school, obtaining directions, or seeing if there's a gas station nearby. You should be aware that when using true GPS, you need to have direct line of sight to the sky. If you're indoors, then GPS may not be the best option for you. That's why "aGPS" (assisted GPS) is starting to become a better option for taking advantage of the location-based services – since you can ask the cellular network to step in when the satellite can't help.

Note: Before you sign up for any service that enables people to see your location, be sure to authorize exactly who can find you. Just to be on the safe side of privacy.

SERVICE: Check with your service provider on rates for accessing GPS, aGPS or Location Based Services.

RATE PLANS

Voice Rate Plans

The term "voice" is used to describe when you are talking on the phone – both receiving and making phone calls. This is most likely the main reason why you have a phone (unless you're using a data device for checking email, for example). You're asked to sign a contract for a length of time, typically two years so you can get a lower price on a phone. When you're looking at selecting a rate plan, there are a variety of factors to consider. Let's review the basic areas to think about when selecting a voice rate plan:

- **Minutes** – You need to think about how often you'll plan on using your cell phone. If it's just for emergencies or to feel safe, then you will want to lowest plan your service provider can offer. If you plan on using it as your lifeline or to run a business, then the higher rate plans are necessary. The first time you're signing up with voice service, your needs may be a bit difficult to estimate and you will have to make a best guess. However, if you're a seasoned cell-phone talker, then look at the average minutes you talked over the past three to six months on previous cell phone bills to get an average of what you normally talk in a month. It's best to skew a bit towards the high side of a minute bundle than on the lower end. Note: If you go over your minutes, you'll be charged a "per minute fee" above and beyond your allotted monthly bundle. If you are under, then you are in good shape. If your actual usage is significantly under or over your allotted monthly minutes, then you may want to consider a new rate plan.

    o <u>Nationwide calling plans</u> typically don't charge you extra for calling long distance within the US, which means it may be cheaper to call cross country on your cell then from your home phone.

    o <u>Roaming</u> is something to consider if you plan on using your phone in places other than your hometown. Check your service provider's coverage area to see if they offer service in the other places cross the US you wish to visit. If not, then roaming fees may apply.

    o <u>Rollover minutes</u> can be a good way to save a few leftover minutes in one month and use them on a rainy day in the future without being penalized for going over your allotted minutes during the later month. It's kind of like depositing minutes into a bank account for later use to avoid overage fees.

- **Mobile to Mobile calling** – Some rate plans allow wireless minutes to be considered free (does not count against your monthly rate plan minutes) when you call another cell phone. Most of the time, this rule applies to those who use the same service provider that you use, but there may be exceptions to this rule depending on the service provider.

- **Pooling plans** – This is best for families who have multiple lines but don't really talk that much. You can purchase a large number of minutes that folks in your family can all use at the same time. This can be an economical way to keep everyone in the family talking, with the benefit of it all being on a single rate plan and bill.

- **Nights/weekends** – Another way to keep costs down is by using your phone between certain hours where you wont be charged your monthly minutes; typically on hours when most people are not using their phones as much. Check with your service provider to find out the hours for nights and weekends.

One thing to note is that most people are tied to their phone number more than they are to anything else. If you have a phone number that everyone knows but you're thinking of switching service providers, rest assured that you can keep your cell phone number and take it with you to another wireless carrier. This way, you can continue to be reached at the same number while you're shopping around for the best deal or better coverage.

Data Rate Plans

Communicating without talking requires "data." There are four main ways to keep in touch and be entertained, and each way has a different fee structure to consider. Let's explore these areas:

- **Messaging (SMS/MMS) charges.** You are billed by the message and can also purchase messages in bulk per month. These fees apply to both text messages and Multimedia Messages, allowing for 200 messages a month, for example.
    - Overage fees may be applied on a per message basis x the number of messages sent over your monthly bundle amount.

- **Wireless internet charges.** This is charged by the amount of data used. Most rate plans are based on kilobits usage. But, really, how much is a "kilobit"? I won't get into a technical explanation here but you will need to determine if you plan on using the web a little or a lot. As a general rule of thumb, check out these examples:
    - <u>Small</u>: If you just want to check sports scores and download a ringtone every now and again, you can get away with the smallest data rate plan.
    - <u>Medium</u>: If you want to check your email or IM, you may be best served looking at a rate plan around in the 5 MB size.
    - <u>Large</u>: If you plan on watching Mobile TV or listening to streaming radio, I would encourage you to have an unlimited data rate plan.

- **Premium service charge.** There may be times that you'll want additional access to new stuff that's not included in your monthly internet service fee. Such examples could be access to HBO content on Mobile

TV or an advanced music service.
- o These charges are typically on a monthly reoccurring basis, in addition to having wireless internet access.

- **Downloads.** When you purchase a ringtone or a game, there will be a one-time fee for that specific download and the charge can be added to your wireless bill or credit card. Due to digital rights management, you will most likely not be able to share the download with others. It will remain on the device that purchased the download.
  - o Be mindful of some downloadable interactive games, for example, that may also require a monthly subscription fee for access to a "virtual community." This may be another example of a premium service charge.

Of course, you'll need to check with a sales rep to be sure you're on the right rate plan for your needs.

If you plan on using data on a PDA or another device to either access your corporate email or to sync your Outlook content over-the-air, you'll need to check with your service provider for separate business/enterprise data rate plans. Most likely they will be different from the consumer data plans that you would use on a basic cell phone, like the ones mentioned above.

Right-sizing your rate plan

Working in the wireless industry, I never had to pay a cell phone bill. It was just a perk of the job (and a really nice one at that!). After 12 years, I ventured out on my own so I could finish this book. Suddenly I faced the reality that I was going to have to pay for my wireless service for the first time. Wow! – I was surprised to feel intimidated by the whole experience. Finally, I understood what everyone else feels when forced to make a decision on the right rate plan. I was able to put my own advice to the test, which proved helpful after the initial shock wore off.

If you've already had wireless service and the ability to see your actual usage from a previous bill, then you were one step ahead of me when I started down this path, because you had a record of your usage that could be averaged out over time. Use that as a guide when "right-sizing" to match up your voice minutes, text messages, and internet usage to find a comparable rate plan that your service provider offers without being significantly over or under in your actual usage. When in doubt, call customer care and ask them for a recommendation. Since I was planning on using my cell phone to conduct business as well as calling my friends/family, I decided the best thing was to sign up on a higher rate plan and then let my normal usage average out. After three months, I called customer care and looked at my own usage to "right-size" my plan to be more economical by matching my rate plan closer to how I was actually using my phone.

A good tip to use when you're not sure of your actual usage is to select a higher rate plan than a lower one. Overage charges have a tendency to get expensive so you should try to avoid that if at all possible. Think of it this way... it's best to plan ahead for a higher bill rather than get shocked after the fact. That said, be sure to revisit your rate plan vs. your actual usage every year to ensure that you're spending your money wisely. You never know, it might be a good way to save a few bucks in your monthly budget that can now go towards a fun purchase!

Prepaid is now cool...

Instead of paying for your monthly usage after the fact, there are options that allow you to pay for minutes up front called "pre-paid." Although you may pay a higher per minute charge for this option, there are some benefits: *No annual contract. No credit check. No overage charges* – basically, no surprises on your bill. It's a great option for kids using wireless for the first time or adults who don't think they'll use enough minutes per month to justify a traditional rate plan.

GLOBAL ROAMING

In the past, the lucky ones who were able to "travel across the pond" (meaning: go to Europe) were thought to be unavailable via their cell phones – they had to rent a phone which meant having an unfamiliar phone and a different number. Quite a hassle when you're on a big trip! What would happen if one of your friends needed advice on shoes and you were not available to help her through this crisis? It would be devastating! (Insert your own example of something tragic if you can't be reached. This just happened to be mine!)

Today is a different world and our cell phones can go with us practically anywhere (check to make sure your phone supports international calling in the countries you plan to visit). That means you can use your own phone and be reached with your own phone number while you're touring Greek Temples and the Pyramids in Egypt. But you need to know a few things before you go...

International rate plan

Quite often, you're not charged long distance when you are in New York and call someone in LA. But when you call from another country, your wireless carrier gets charged a fee and they pass that along to you in the form of a premium per minute fee. To avoid any extra charges, you can sign up with your carrier's international rate plan. This will reduce the fees you pay when traveling abroad and want to call home. In some situations, you may be restricted from making any calls internationally unless you have this rate plan, so check with your wireless carrier for more details before you leave.

To dial back to the USA – Use the "+1" rule

Once you get to your exotic destination, you might find calling home to be a challenge – unless you know about the +1 rule. While abroad, you're required to dial the country code for the United States before dialing the normal phone number when you want to brag about your adventures. To do so, find the "+" on your phone keypad (commonly found on the "0" key), then "1" and dial the rest of the phone number as normal.

Checking voicemail while abroad

When you press and hold "1" to access your voicemail while in the USA, you might not realize that it's actually dialing a phone number. Then, based on detecting your cell phone number, you are routed to your voicemail box. This phone number is a USA-based number and will not be accessible when you're abroad without remembering the "+1" rule. You can either: (1) go into your phone and update the phone number you automatically dial by adding the "+1" or (2) you can dial your own number (still using the "+1" rule). Both will work to retrieve your voicemail.

Last minute reminders – International power chargers

It's easy to forget something when you travel, like your toothbrush. But don't forget your phone's power charger. And while you're at it, get an international power adaptor for your phone's power charger. You went through all this trouble to make sure you have a phone that works and you added the international rate plan. You even know how to dial back home. It would be a shame after all that to then have your phone battery die with no way to plug it into the wall to recharge.

Some phone manufacturers like RIM offer a variety of international plugs that come in the box. Others will sell them online as an accessory. You can also use any international power adapters/converters that you can buy at travel stores. Be sure to check the power outlet style before you leave on your trip. Otherwise, you may just need to buy an international power charger once you get there.

Last Resort – Rental phones

If you don't have a device that will work internationally, it's still possible to rent a phone in the country you plan on visiting. Check with your service provider to see if they have a program that can offer you discounts and a company referral so you can still be accessible while on your trip.

Part 3
Explore Devices

Earlier, you learned there are different levels of the network and you saw that they can be divided by speed. Devices share the exact same ladder structure that shows how fast you can access data over your phone. So if you see a device that says it's EDGE, then you know the average speed it can access is 100 kbps over the internet, similar to your cable modem at home.

If the network is like a helium balloon, allowing the ladder to grow higher and higher over time as it increases in speed and capabilities – then devices are like rocks, starting out as high as they can on the ladder and either stay there or fall. Meaning, devices can never go higher than the maximum speed they are originally assigned, even if the network is capable of more. But devices can "step down" to a lower level on the ladder if necessary or if the higher network level is not available in that area.

Remember when your mom said the age-old phrase of "Don't judge a book by its cover"? This can also be true when applied to devices. If you see a device offered as 2G, you already know it will start at 2G and never go any higher. But sometimes a next-generation device can LOOK exactly the same, but it has newly increased functionality and now that same looking device can access 3G. If you check closer, you'll see that although it looks the same, the device's model number has changed. This is your clue to know that the manufacturer used the same form factor but improved its functionality, which in this case was giving it higher access on the network ladder. This doesn't happen often – but when it does it can be confusing, so it's worth a mention here.

Let's set up our ladder again.

Now it's important to know that devices and the network work together as a team when it comes to supporting different services.

I'll place pictures of devices next to our network acronyms on the ladder to help give you an example of putting your newfound knowledge to work. It's important to understand that the devices I've chosen are just samples as they change quite frequently, so don't think that these devices are your only options. This is merely a tool to help emphasize a point:

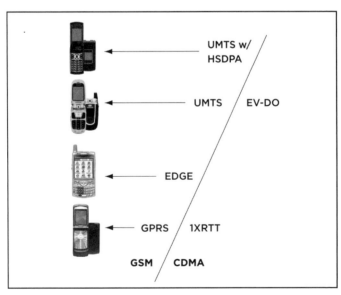

WHAT'S YOUR FAVORITE BAND?

I'm not talking about the type of band that might have won a few Grammy awards and has a new CD coming out – I'm referring to bands of "spectrum" that exist in the sky. The FCC has allocated various frequencies of spectrum (layers in the sky) into bands (specific set of frequencies grouped together) so TV does not interfere with radio, which does not interfere with wireless, which does not interfere with HAM radios, which does not interfere with garage door openers and cordless home phones. You get the picture. Now we'll consider this band… "Mega hertz" or MHz for short.

Regarding MHz, let's look at the frequencies that cell phones can support. In true wireless fashion, there are different names to review here. Luckily, these names are more self-explanatory than our network acronyms.

- **Single Band**: If you have a single band phone (850 MHz), its most likely analog – and it's time to upgrade! We really need to talk…

- **Dual Band**: If your phone supports two frequencies (850 / 1900 MHz) then you have a "dual band" phone. This is great for roaming across the USA but won't get you far "across the pond." This is also what is most commonly supported with CDMA handsets.

- **America's Tri Band**: If your phone supports three frequencies (850 / 1800 / 1900 MHz) then you have a good start with which to roam. It will get you everywhere in the US, but you'll only be able to use it in just a few places around the world.

- **International Tri Band**: This phone also supports three frequencies (900 / 1800 / 1900 MHz) and you can use it around the world, but you didn't buy it in the US. Reason being is that you are missing the 850 MHz frequency which is extremely important to have when using a phone here in the United States. End result? You may have a great device to take with you for global travel, but it will be of limited use to you in the US.

- **Quad Band**: If your phone supports four frequencies (850 / 900 / 1800 / 1900 MHz), then you truly have a global phone. This device will serve you wherever you need to go.

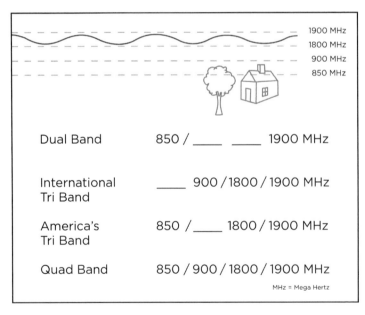

Dual Band	850 / ____ ____ 1900 MHz
International Tri Band	____ 900 / 1800 / 1900 MHz
America's Tri Band	850 / ____ 1800 / 1900 MHz
Quad Band	850 / 900 / 1800 / 1900 MHz

MHz = Mega Hertz

To see which frequencies your phone supports look on your box, refer to your user's manual, or look up your phone online. Also, your sales rep should be able to let you know. When shopping for a phone, make sure your sales rep knows what counties you might visit to make sure you get the right phone to meet all your needs.

Note: You may have seen the term "GHz" in relationship to bands. This stands for Giga Hertz, and you can use this term in place of MHz by doing nothing more than moving a decimal place over one digit, so a phone that is 1900 MHz can also be referred to as 1.9 GHz. Sort of like saying you can have 100 pennies or 1 dollar. They both have the exact same monetary value, but are converted differently.

If you have ever tried to use your phone in another city – or another country for that matter, all of the sudden you realize these bands have meaning. North America supports two bands: 850 MHz and 1900 MHz. The 850 MHz band was the original band used for our 1G or analog networks. As time went on and new wireless carriers wanted to enter the playing field, the FCC said, "Hey there – we don't have room for you on 850 MHz. You're going to have to get your own frequency." OK – so they did. Ever heard of "PCS"? The term PCS was given to identify the newly created band of 1900 MHz. Just so you know... PCS is nothing more than a brand name of a band width. An old boss of mine told me that back in the mid-90s. I guess it stuck. Sprint made it famous by trying to differentiate themselves by saying they are the "All digital PCS network." It didn't mean they were any better than anyone else with their digital technology, it just meant they didn't start with Analog roots. They started with Digital or 2G and only offer service on 1900 MHz.

As the wireless industry continues to grow in functionality, it's always possible that even more bands will continue to open up. For example, future Mobile TV applications may take advantage of the 700 MHz band for room to enable real time streaming video while not impacting today's network used to make and receive phone calls.

WHAT ARE YOUR OPTIONS?

Every person who buys a phone may be looking for something different, but certain themes have evolved over time and maybe you can relate to one of these categories. These themes may also help to narrow down the playing field once it's time to select a device when you have several to choose from.

WAYS TO SELECT A PHONE	DESCRIPTION
Price	"I'm on a budget"
	Many people are on a budget and can't afford an expensive phone. I mean really – It's just a phone. You talk on it. The good thing to know is that these days cheap does not mean basic. The costs are coming down on phones, and these days, you can get a really good camera phone for under $50. Also, you can check out BOGO (Buy One Get One) offers which bring the price down per handset. Also, check out the devices that have rebates. You may spend a bit more at the register, but in a few weeks you have some bucks back in your pocket.
Feature	"There are a few things I want to do on my phone."
	There's a ton of features & services that are available and new ones come out every day. Some of these features, such as cameras and Bluetooth, are more popular and more widely available. Services such as Push to Talk (Walkie Talkie style a la Nextel) are really cool to use but are not available on every device. Some services such as streaming video or music downloads are only available on devices that work on the high speed data networks (3G), so you need to know what you're looking for to be sure your phone can support what you want to do. Be sure to ask and try it out before you buy so you won't be disappointed later.
Fashion	"My phone says something about me – it's got to look good and show off my individual style."
	Thin is in and phones that reflect this trendy form factor fly off the shelves and often command high prices when they first hit the streets. Another trend comes in the form of colors… and then the fashion frenzy starts all over again. It's one of those "gotta have's" to complete the high-style look.
Productivity	"I need a PDA to stay organized - I have to respond to email and get real time access to my calendar."
	Certain devices offer business applications such as corporate email access and synchronization to your Outlook calendar - which is critical in a busy professional's life. Also, there are folks that don't want to lug a laptop around all the time, so instead they would rather access specific information on something a bit more portable. They keep business moving at the speed of technology.

Sometimes, a wireless carrier will work directly with the handset manufacturer and will put their name and logo on a device. Folks have often mistaken this to think that Verizon is a manufacturer of cell phones. This isn't true, but they did work closely with the real handset manufacturer to create something special for their customers – including customized menus and icons related to the phone's features & services. Cingular/AT&T, Verizon Wireless, T-Mobile, Sprint/Nextel… they all offer their own unique customizations on phones. The term given to the experience of interacting with the menu, icons and the different steps you take is referred to as "UI" (User Interface). There are experts who do nothing but study people's behaviors regarding how they use a phone, and then they make recommendations for designing the software and menu structure to be the most efficient and easiest to use. The goal is to offer as much functionality as possible with the least number of steps (also known as "clicks") to get there. Hopefully they have succeeded in their mission and you find your phone easy to use. If not, then you may want to look at having your next phone from a different manufacturer. The reason? Well, the logic that's built into one phone is often carried over into other phones by the same manufacturer for a consistent experience.

Form Factor

This refers to the way a device looks. It also can refer to how the device opens; for example, does it fold in half or not? Often, there is more than one name for the same design, so I'll try to list as many as possible. Using these terms is a good way to communicate your preference to a sales rep when picking out a phone.

Cell Phone Decoder Ring

DEVICE FORM FACTOR	DESCRIPTION	IMAGE
Candybar or bar phone	This phone is long and narrow, and it does not change its shape.	
Flip phone or clamshell	This phone folds in half to protect the keypad.	
Slider	This phone is like a candy bar that rises and collapses on itself.	
Jackknife or swivel	This phone is similar to a candy bar design, but it rotates to the side on a hinge to make a larger candy bar.	
Dual Flip or Dual Hinge	This is a device that looks like a normal flip phone, but you can open it a second way to take advantage of a different layout of the same keypad – now in QWERTY mode.	
PDA with a full QWERTY keypad	This is a PDA that uses a full key board. It's wider in size than a candybar.	
PDA with a SureType keypad	This keypad is a modified QWERTY that uses one key to represent two letters of the alphabet making the device look more like a candybar phone and less like a PDA.	
PC Card or PCMCIA	This credit card-size device enables a laptop to have access to the wireless network.	

PHONES

Now it's time to talk about the fun stuff!

Handset Manufacturers

When people first think of a cell phone, they wonder who makes it. "Is that a Motorola?" This is one of the most recognizable names of handset manufacturers, and they are a good old US based corporation that has been manufacturing phones for many years – and so have others. Below is an overview of the larger handset manufacturers who sell their phones in the US:

- Motorola: US-based handset manufacturer that designs devices that are very fashion forward. They are the creators of the "thin is in" trend when it comes to handset form factors.

- Nokia: This giant gave cell phones a name back in the mid-90s as a handset manufacturer out of Finland. They're strong in terms of providing devices that are high in functionality and a powerhouse in providing the latest features.

- Sony Ericsson: Another global giant, this one out of Sweden, also has features and functionality on the mind when designing phones. They are strong when it comes to offering their music line of devices together with Walkman, a fellow Sony brand.

- Samsung: This Asian manufacturer was quick to adapt to handset trends such as providing very fashionable and thin clam shells that look great and also work well.

- LG: A fellow Asian manufacturer that appeals to all the same great attributes Samsung has been able to capitalize on in recent times by offering nice form factors and good quality.

- Pantech: They are new to North America, but have been very innovative elsewhere around the globe for years by being one of the first manufacturers to put a camera into a phone before we thought it to be possible here in the US. Imagine what they'll do next!

- palmOne: The makers of the TREO and Palm software have since diversified and opened themselves to also offering Windows-based PDA devices as well.

- RIM: The name of this Canadian-based company actually stands for Research In Motion, but they are better known as the creators of the Blackberry for corporate email access. They continue to prevail as the leaders in providing secured access to corporate and personal email on devices that can be found everywhere from Hollywood to board rooms.

- HTC: This manufacturer may not be a household name but you've seen and heard of the products they've made in the past. Now they're coming direct to you as a handset manufacturer. They predominantly provide Microsoft Windows PDA devices with nice form factors and often a choice of a hidden QWERTY keypad.

- UTStarcom: They offer stylish and smart devices in a range of form factors and prices to give you a full glance of a variety of options.

- <u>Kyocera</u>: This company, which focuses in the CDMA space, offers rounded form factors that are just as attractive as they are fun to use.

- <u>Apple</u>: New to the area of offering cell phones, the Apple iPhone offers compatibility to your laptop with a rich music experience, while offering a new way of interacting with the phone menu, thus setting a new standard for software and hardware design.

Of course, there are many more handset manufacturers available than what is listed here, but it's just to get you started...

Did you ever wonder how some of these companies got their names? Sometimes, they are a combination of two companies that merged together like Sony, who had manufacturing experience of electronics and Ericsson, who had wireless experience – to create Sony Ericsson. Other times, they are the name of the company Founder, and such is the case with HTC (owned by Mr. H.T. Cho). LG is actually short for Lucky Goldstar which also makes TVs and appliances. My favorite name origin is for Nokia, believe it or not. They were formed about 100 years ago in Finland as a logging company. They would cut timber and float them down the Nokia River in Finland to the sawmills. It was only 90 years later that they ventured into one of many various businesses until they struck gold... in the form of wireless. They sold off the remaining companies not related to telecommunications and focused on their new core business, quickly becoming one of the most recognized brands in the world.

SMARTPHONES/PDA

Perhaps you have been thinking about a device with a bit more substance – where you can keep track of your calendar, emails and open Microsoft Word and Excel documents all on your phone. Enter the world of "data devices" (or PDAs) where the operating system is just as much the point of differentiation as the form factor. Just so you know… data devices are not just for business users; they're also great for college kids, soccer moms or anyone who likes to stay organized and doesn't want to carry multiple devices.

Check out the chart on the next page that shows how to evaluate a data device by the operating system. The devices may change over time but it gives you a good idea what to expect.

OPERATING SYSTEM	DESCRIPTION	IMAGE
BlackBerry	Email: Blackberry Personal and Business Display: No touch screen Keypad: QWERTY or SureType Apps: View text only (no graphics) Media Player: Newly available Camera: SureType only	
Microsoft Pocket PC	Email: Personal and Business Display: Touch screen Keypad: Touch screen or QWERTY Apps: View, edit & create attachments Media Player: Windows Media Player Camera: Optional	
Microsoft Smartphone	Email: Personal and Business Display: No touch screen Keypad: Phone keypad or QWERTY Apps: View-only attachments Media Player: Windows Media Player Camera: Yes	
Palm	Email: Personal and option for Business Display: Touch screen Keypad: Touch screen or QWERTY Apps: View attachments Media Player: Real Media Player Camera: Yes	
Symbian	Email: Personal and option for Business Display: No touch screen Keypad: Phone keypad or QWERTY Apps: View attachments Media Player: Available Camera: No	
Apple iPhone	Email: Personal Display: Touch screen Keypad: Touch Controls Apps: Sync with MAC Media Player: iTunes Media Player Camera: No	

What's the IQ of a Smartphone?

The term "Smartphone" can be a bit confusing, and it can have a variety of meanings, depending on who you ask. Here are a few commonly used references to describe a Smartphone, as many handsets want to be a part of this prestigious category:

- Smartphone #1: A device that is considered to be a PDA (Personal Data Assistant). It stores a copy of your Outlook calendar, email and other organizational items. Traditionally, this device has a large screen; either a full QWERTY keyboard or Touch screen for text input, but it's also a phone. It may support one of several Operating Systems which have names like Palm, Pocket PC and even Blackberry. This group is more commonly referred to as "data devices." They may look intimidating at first glance, but they act just like mini laptop computers that also allow you to make phone calls.

- Smartphone #2: The proper name of a Microsoft Operating System that allows access to Outlook, email and other Microsoft applications on smaller and more traditional-looking phones. It is a more basic version of the Pocket PC operating system that does not include a touch screen.

- Smartphone #3: A device that combines multiple devices into one – such as a camera, MP3 player and phone. It may also include a full QWERTY keyboard for messaging but lacks the PDA functionality. My personal opinion? – This is a stretch to include this description under the Smartphone category as it does not offer any laptop functionality that is usually associated with this term, but it is important to mention to avoid further confusion.

As you can see, this term encompasses a lot. They all must be pretty smart!

QWERTY

I'm amazed by the amount of people who know the term "QWERTY" to describe a full computer keyboard, but the light bulb usually does not go off until I ask "Do you know why?" Look at the top row of letters on your computer's keyboard from left to right. What does it spell? Q.W.E.R.T.Y. Don't believe me? Try typing it!

Touch Screen

An alternate means of text entry, other than using a keypad, is to have a device that can accept commands by the touch of your finger or a pointer called a Stylus. Touch screens are traditionally found on PDAs and provide a nice, clean look to the device as well as a large screen to work on.

WIRELESSLY ENABLE YOUR LAPTOP

PC Cards

This credit card-shaped device helps your laptop to connect to the internet (over the wireless network) without the need to plug into a wall for DSL or dial-up usage. A PC card is a great alternative for those with older laptops who don't have built-in WiFi access (or when WiFi is not available) and want freedom to move around while staying connected. "PC card" is short for "PCMCIA." Someone once told me that the acronym stands for "People Can't Memorize Cellular Industry Acronyms." Actually, it stands for something else – you get extra credit for knowing what it really stands for without looking at the Glossary!

Getting set up

When you buy a PC card, it will come with the card itself (sometime you have to plug in a separate antenna) as well as a CD ROM that you install in your laptop. An installation wizard will pop up walking you through downloading some software so your laptop knows how to talk to your PC card. When the two work together, you will have established a wireless connection to the internet (think of it like wirelessly enabling your DSL line – so you can surf the web while sitting in your backyard). An icon will appear on your desktop so when you're ready to get going, all you have to do is (1) put your card into the Type II PC Card slot (this is the little door that reveals a slot on the side of your laptop), (2) launch the PC card application on your desktop and (3) start surfing.

Selecting a PC card

If you want to sound smart to the Geek Squad, ask for "a PC card that is compatible with your Type II PC card slot in your laptop." They will look at you with a renewed sense of excitement and enthusiasm as if the human race has finally caught up with them, regardless of the fact that you read it in this book. You are learning, and that's what matters! OK, follow up question – they will ask you "which one would you like?" *OH NO!* You think. *I have to choose?* Yes, sorry to say. There are as many different cards just as there are different cell phones. You might recognize some names such as Sony Ericsson, and others you may not recognize such as Novatel or Sierra Wireless. Be sure to find one that is higher up on the network ladder to ensure you are getting the fastest speeds available. Other cards also include WiFi as well.

Your PC card is actually a wireless device

The card itself is similar to a cell phone in that it requires activation on a wireless network and can only work when you have good signal strength. (Your signal strength can be found when looking at the software running on your laptop) The good news about this card is that you never have to worry about charging it since it works off your laptop battery!

There are basically three things you need to think about when selecting a PC Card that's right for you:

1. <u>What wireless carrier do you want to work with</u>? Rate plans may differ – so shop around. You also need to be sure you will have coverage in the areas you want to travel.

2. <u>How are you planning on using the internet on your laptop</u>? Is it for downloading email and connecting back to your corporate server? Or will you be streaming video and downloading music? This will help you to decide where in the network ladder you will need to be. The higher the better! For a better understanding of the various network speed definitions, refer back to the network section.

3. <u>Do you actually need a PC card or are you open to other alternatives</u>? Some newer laptops include a wireless module (separate from WiFi) that allows access to a wireless service provider without the need to buy an extra card. For a simpler solution, you can also tether your phone to your laptop by using a data cable – thus using your phone as a modem.

Centrino

This is the term to indicate if your laptop has WiFi built-in. If so, then you can access the WiFi internet. You may still have a PC Card slot for access to the wireless network when WiFi is not available. For more information on WiFi, refer back to the Services section of this book, find the Advanced Comfort Zone section for a chart that explains the various versions of WiFi access.

Laptop Connect Service

I am using this term to represent the rate plan that is designed specifically for PC Card access over the wireless internet. I would strongly encourage an unlimited data rate plan, which is separate from the internet rate plan you can get for your cell phone.

Wirelessly Enable Your Laptop

OPTIONS	CELLULAR OPTIONS	WiFi OPTIONS
Use Existing Cell Phone	• Phone as a Modem • Tether Cable or Bluetooth • Laptop Connect Service	• N/A
Use PC Card	• PC Card • Laptop Connect Service	• PC Card or • Laptop with Centrino
Built In	• SIM Card for Imbedded Module • Laptop Connect Service	• Laptop with Centrino

ACCESSORIES

Accessories are a way to enhance the experience of your device – whether it allows you to carry your phone on your hip, talk hands free with a headset, or decorate it with stickers. When selecting an accessory, you'll need to know the make and model of your phone to ensure compatibility. If you don't remember, then take note of the plug on your power adaptor that connects with your phone, since that play a part when selecting several accessories.

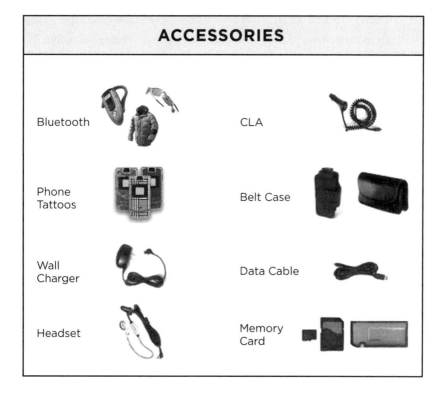

Bluetooth

You learned earlier in this book what Bluetooth is all about – it's a technology that allows your phone to communicate with another device without wires. There are many uses for this technology, but the most popular is to connect to a Bluetooth headset (remember: your phone needs to support this technology). Bluetooth headsets are becoming more creative by including the technology into sunglasses, jackets and other apparel items. Many new cars also include this technology through a speakerphone system. Additional uses for Bluetooth include file transfer from a phone to a Bluetooth compatible printer or listening to music over Bluetooth stereo headsets (this profile is called "A2DP").

Cigarette Lighter Adapter (CLA)

Also referred to as "Vehicle Power Adaptor" (VPA), the power charging accessory is a must-have for any wireless user who talks more than what their battery may allow and is frequently in their car. The CLA connect to your car's charging port similar to plugging the phone into the wall for charging.

Phone Tattoos

When you're sitting at a table with a few of your friends and you all have the exact same phone, how can you tells yours apart – especially when you can't hear your ringtone? Or… how do you plan on matching your phone to your outfit without having a collection of one in every color? Enter "phone tattoos" – the latest accessory to personalize the outside of your phone – stickers that come in all different shapes, colors and some that even include some nice sparkly bling. Swap them out with the seasons or as your sense of style changes.

Belt Case

Often our lives have us on the go a great deal, and although we need our phones with us, sometimes it's not convenient to put it in our pocket or purse. Therefore, a belt case is a great functional necessity and can be a fashionable option as well. A fitted pouch holds your phone and fixes itself to your belt, providing easy access for answering when it rings.

Wall Charger

This necessary power accessory, the phone wall charger, plugs into a standard outlet and comes included in your phone box. It will fully charge your device in about an hour. Additional chargers are available for purchase, so you can have one in your office, at home and one that is ready for travel.

Data Cable

This cable, which has a laptop USB port on one side and your phone's power connector on the other, is a great multipurpose cable. You can do several things with your laptop and phone together by using this cable:

- Move info from your phone to your laptop (for transferring pictures).
- Sync info from your laptop to your phone (such as copying your Outlook so it shows up on your phone or PDA) or transfer music (otherwise known as "Sideloading") from your library to the MP3 player on your phone.
- Use your phone as a modem for internet access on your laptop.

Corded Headset

Many states have implemented a "hands free" driving law which indicates that you can not have your phone up to your ear while driving. When this is the case – and you still want to be available on your mobile – purchase a corded headset. It's also a convenient way to talk while typing on your computer or carrying groceries into the house – especially when your phone is not Bluetooth compatible. Typically, headsets include a single ear bud but stereo headsets have dual ear buds for surround sound music as well as talking.

Memory Cards

Some of the newer phones include a place to put a memory card which will store photos, games, music or anything you would like to save when the memory on your phone is not enough. This card is similar to storing computer files on a CD ROM, but it's for your phone. They are found either under the battery or on the side of your phone. Memory cards (also known as "external memory" or "removable memory") may also be accessed from the side of your phone and called "hot swappable" – because you can insert a new card without having to turn your phone off.

- <u>Various types of memory cards are available</u>: There is a variety of types (form factor) & sizes (memory capacity) and may vary depending on the handset manufacturer. The most common brands are:
    o Memory Sticks from Sony
    o SD cards from Scan Disk

Within each of these brands, there are a variety of sizes such as Memory Stick Duo or Micro SD. There is a range of storage amounts available as well. The card, regardless of the type, will indicate how much it can store, based on

"MB" (mega bytes). Sizes will be indicated in numbers such as 64MB, 128MB or 512MB, 2 GB and beyond.

Memory cards have metal connectors on them, similar to those on batteries, which require that they be inserted in the proper direction. Be sure to watch for this while inserting an extended memory card.

Part 4

What cell phone should I get?

Hint:
Apply what you've learned...

Many people like to ask the most difficult question first: *"What cell phone should I get?"* It may seem like a simple enough question to you, but in reality the answer can change with each person. Meaning, in order to know which device to get, you need to know *how you want to use it.* The good news is that, in your mind – whether you know it or not – you probably have already started answering the question for yourself.

There are three parts to answering this question, and they need to be asked in a certain order so you can find out what's right for you. It should sound familiar because we're now going to apply the basics you've learned from reading this book so far.

ASK YOURSELF THREE SIMPLE QUESTIONS...

Yes, it's now time to ask yourself some questions, and the first one is…

Network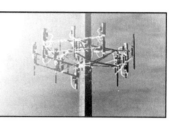

#1. WHAT SERVICE PROVIDER SHOULD I USE?

- Who has the best coverage?
- What digital technology do they use? (Think of the network ladder)

The first thing to think about in selecting the right cell phone is: Which service provider offers the best coverage in your area? If you skipped the beginning of this book, then you missed the fact that this is truly the corner stone of creating a wireless solution and how you answer this question will ultimately determine which devices are available for your selection. This company, often referred to as a service provider or wireless carrier, will be the one to make sure you can answer and place a call wherever you go.

There are national service providers known as Cingular (AT&T), Verizon Wireless, Sprint/Nextel and T-Mobile. There are also regional service providers including Alltel, U.S. Cellular, Dobson, Cricket and MetroPCS; and local service providers like Cincinnati Bell. In order to answer the question of "Which service provider should I use?" – First you need to acknowledge if you already have service and are locked into a contract. If you have a current contract, it may extend as long as two years and could be costly to break. If you're already under contract and get good service, then you should consider staying unless you feel that you don't get the best coverage possible. Coverage is nothing more than just making sure that your phone has the ability to make/receive calls in the areas you visit, such as where you live, where you work, and where you play or run errands. If you don't have good service (meaning – you can't make a call and you don't have any signal bars on your phone or your calls drop frequently) in any of these areas, then you should consider switching service providers.

Birth of new player - MVNO & Resellers

In the land of opportunity that embraces competition, a new breed of wireless companies has hit the scene. The term given to this business is called "MVNO" (Mobile Virtual Network Operator) or better known as "Reseller" – which is essentially a business that buys minutes from a larger wireless service provider and will sell the airtime back to you under a different name. You may be asking yourself, "Why bother?" Well, I say, "Why not?" Cell phones are an expression of our individuality, personality and style. They say something about us – we are trendy, we are practical, we like technology and gadgets, we like Mickey Mouse – or whatever you want to say. Then, the stuff we do on our phones says even more. The personalization of your experience is exactly what the MVNO/resellers are going after. They offer a complete package to you: unique phone, cool ringtones, content on the web, etc. – allowing you to become a member of a certain club or supporter of a particular group. They have created a niche for those that love sports, music, Disney characters, social networking/blog sites, etc., and they enable your wireless experience to reflect that passion.

- Disney Mobile: Disney Mobile launched with their themed experience that's good for the whole family. That means Disney magic is mobile!

Mickey Mouse characters guide you through the phone's menu as you select *The Lion King* song "Hakuna Matata" as your ring tone. Play a driving game with the characters from the Disney Movie *Cars*. There's other unique content such as having the ability to interact directly with Disney Radio by dedicating a song from your phone.

- <u>Helio</u>: Perhaps you want something that's a bit more high tech with unique handsets that totally talk your language! Helio, another MVNO, is all about keeping it real by offering access to MySpace and other cool social networking sites from your phone and gettin' your groove on with video ring tones.

- <u>Amp'd</u>: Another totally cool service that's all prepaid for those that are young at heart is Amp'd Mobile. This extreme sports, Hollywood-loving, in-your-face wireless service provider offers rate plans with names like "end the boredom" and "unlimited forever overdose." You'll be sure to feel like a VIP while you're hangin' in wireless style.

Others like P. Diddy have been looking into offering similar services – his thought is to be all about the music and a cool lifestyle. Stay tuned!

Converged Services & 3 Screens

This section is especially for those of you who are asking yourself, "What exactly are the three screens that everyone keeps talking about?" They are TV (by offering cable), laptop (by offering internet access), and your cell phone (by offering wireless service) – and sometimes all on one bill.

You may notice that a variety of communication companies are starting to offer multiple services. When your cable TV provider offers wireless service, they may be doing so as an MVNO/Reseller. Other companies, such as AT&T or Verizon, can offer you all three of these services directly.

#2. WHAT DO I WANT TO DO WITH MY PHONE?

What's within my comfort zone?

- **Basic** (Check voicemail)
- **Intermediate** (Send text messages)
- **Advanced** (Watch videos)

Many people like to rush into the third step of selecting their phone, but I urge you to take a little time and think about your Comfort Zone. Will you be sending text messages or need to access email? Do you plan on using a camera or listening to music? Will you be downloading stuff? (Remember what I talked about earlier – Basic, Intermediate and Advanced? It starts to apply here!)

Taking the time to think through this second question will help you determine what features & services you really want (be sure to note how your chosen service provider offers these services) and will ultimately ensure that you get the right phone to meet your usage needs – without getting something too powerful or being under-served.

Once you have determined what you want to do, you can match your Comfort Zone to the network ladder (be sure to take note of the technology your chosen service provider supports) to determine how high you need to climb. Remember the 2G and 3G acronyms we learned about earlier? This will give you direction as we lead into Question #3....

#3. WHAT'S IMPORTANT TO ME WHEN SELECTING A DEVICE?

- **Price** (I'm on a budget)
- **Features** (I want to do certain things)
- **Fashion** (I want to look good)
- **Productivity** (I need to stay organized)

Will I need to travel abroad with my phone?

By the time you are ready to select a device, you may notice that the second question narrowed the playing field and reduced your consideration set. If you are faced with a choice – all of which seem worthy, use Price, Feature, Fashion or Productivity as a way to select from what's left over.

There are times when Price, Feature, Fashion or Productivity drives the decision from the beginning. If you are one of these people who has "cell phone envy," just be sure to go back and review which features & services from within your Comfort Zone are available on this cool new toy, to be sure it really is what you want. After all, you're the one who has to use it.

Note: It's perfectly acceptable for a man to consider fashion as criteria for selecting a device. However, if it makes you feel better, you can substitute "fashion" for "form factor."

As a recap, you'll want to consider the following high level checklist before selecting a device on your own:

- **Which carrier do you want service with?** Certain service providers only offer certain devices. Make sure you're selecting a device that's compatible with the service provider you want to work with. Remember, the technology your phone uses needs to be a match with the technology your service provider offers.
    - o Example: Selecting a service provider will determine the device technology required (CDMA vs. GSM vs. iDEN) as shown on the network ladder.
- **What do you want to do with your phone?** Make sure that your favorite features & services within your Comfort Zone are available when

looking for your ultimate device. It doesn't make sense to get a phone that doesn't support your needs. Be sure to look at the techs/specs of a device to see if the features & services you want are available, then match it up to the network ladder to ensure you have the appropriate level of internet speed for the best experience on your phone.
- o Example: If you want to check email, be sure to select a device that supports POP3. If you plan to respond while on the go, you may need to consider a device with a QWERTY keyboard. Email access can be fully supported on a 2G device.
- o Example: If you want to watch Mobile TV, then you'll need a device that supports this feature. Make sure it's 3G to have the best streaming video experience.

- **Do you plan on traveling internationally?** Be sure to look for a phone that will work in the countries you wish to visit.
  - o Example: Look for a device such as Tri band or Quad band.

- **What's more important to you in making a final handset decision?** There are ways to further narrow down your choices if you have more than one device that supports the features you want to access.
  - o Example: Think about what's most important to you when it comes to making a final decision. Are you leaning more towards Price, Features, Fashion or Productivity?

Remember that your answers may lead you to one or more devices that you can select from, which may be very different from those chosen by your friends or co-workers.

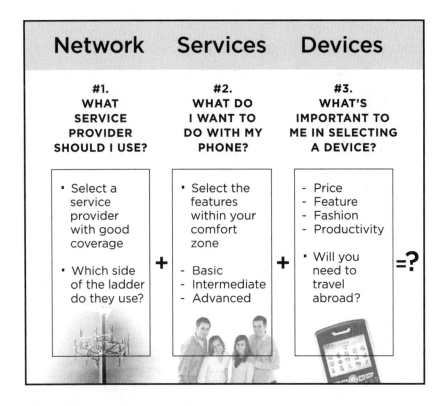

With the major questions answered in the appropriate order, you'll wind up with a positive experience that allows you to have a wireless solution that meets your needs. As I said earlier – "What phone should I get?" is a difficult question because it's about more than just one person's suggestion. It's a custom solution made just for you!

USE A "WIRELESS WORKSHEET" WHILE SHOPPING

Feel free to use this worksheet while you are exploring your options. Use the internet to help you do your research and learn what's available in your area.

You can download a copy of this worksheet from
www.CellPhoneDecoderRing.com to take with you while shopping.

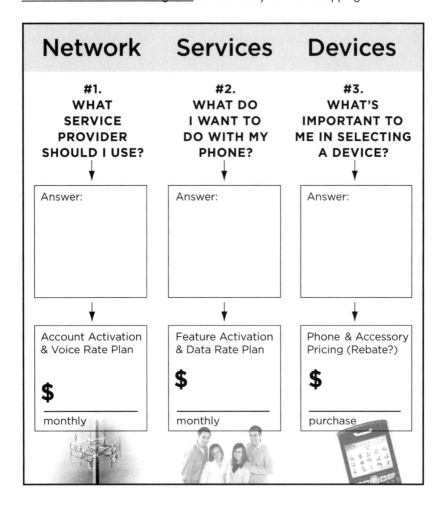

Part 5

Living in a Wireless World

You've cracked the code... now what?

WIRELESS ETIQUETTE (A.K.A. THE RULES OF THE GAME)

Have you ever been out to dinner where the mood is right, the music is soft, you're feeling good... then all of the sudden, the most obnoxious cell phone ringer comes on, some guy answers the phone loud enough for everyone in the restaurant to hear every word he's saying, and then even worse, as the guy keeps talking at top volume, he doesn't even get a clue that others are getting really upset? You just wish he'd shut up!

Cell phones are great, don't me wrong. But when you come across loud talkers at the wrong time, it reminds me that every now and again, we could use some advice from Miss Manners on social etiquette – wireless style.

Here's a question I was asked on this very topic...

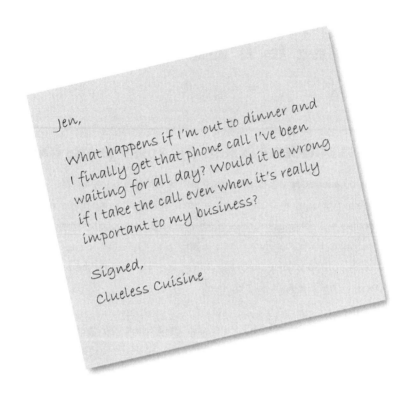

Jen,

What happens if I'm out to dinner and I finally get that phone call I've been waiting for all day? Would it be wrong if I take the call even when it's really important to my business?

Signed,
Clueless Cuisine

Dear Clueless,

Few things are important enough that they warrant disturbing others around you. Answering a call at the table can be considered rude, especially for the ones you're dining with. After all, they took the time to visit with you in person. It's OK to look at your phone to check Caller ID, but send the caller into voicemail and call them back later. Be sure your ringer is set to a low level or vibrate so we don't hear your favorite rock band ringtone over the lobster bisque.

You could always respond to a call by sending them a text message back saying you're "at dinner – will call later." Keep your message brief and don't get into a back and forth conversation. That can be just as rude as answering a call at the table.

If you must, politely excuse yourself from your guests and leave the table. In fact, leave the dining room and enter a common area where you can begin your conversation. Tell the caller to hold while you leave the room or let them talk first while you make your escape. Be sure to watch where you're going so you don't knock over a waiter while you exit!

Always remember that your voice carries and you're not in your living room or office. So be careful not to be too loud. We really don't want to hear what you're talking about. Your conversation doesn't interest us in the least.

Make sure you keep your conversation brief and to the point. Since you left your guests hanging back at the table, they're probably waiting on you before they can start the next course. So finish what you need to do and get back to that table!

If you're the unfortunate guest of someone who violates the "no talking at the table" rule, it's perfectly acceptable to let them know it bothers you – as long as you're polite when doing so.

Here are some other helpful hints that Miss Manners would be proud of:

- When you're watching a movie, it's only natural that you become absorbed in what's happening. A phone that's ringing will quickly snap you back to reality unnecessarily. Either turn your phone off or change the ringer to "silent" when in a movie theater. By the way, "vibrate" is just as annoying as hearing your ringer since it still makes a sound. We can still hear it.

- If I have to hear your ringtone, at least make it a good one. Lousy ring tones are very annoying and can be embarrassing for those who selected it. There are plenty of good ones out there to download.
- Don't leave your phone to ring unattended. You may not hear it when you step away, but we do. And we know how long it rings as well – in fact, I've started humming your ringtone as I fall asleep at night. Save those around you by keeping your phone with you or turning the volume to low so it won't bother folks when you're not there to answer it.

For fun, you can issue a cell phone citation, a pad of paper that looks like a traffic ticket for wrongful cell phone violators. You can find the "citations" online or at novelty shops. Have fun while making your point – and avoid being rude yourself!

FASHION DO'S AND DON'TS

Welcome to the Tech Fashion show, where we will showcase a collection of what's functionally smart or fashion suicide...

DO

- Fashion-conscious folks can coordinate their cell phones with an outfit... and change the cell phone's look based on the season, for that matter.
- Keep your phone fully charged. Nothing ruins a perfectly good image like the low battery warning.
- Show off your latest gadget by placing your phone on the table in front of you... in case someone calls, of course. If your phone is bulky, ugly or otherwise old, keep it hidden in your handbag, backpack or briefcase. You didn't really need to get that call!
- Be a savvy shopper and ask for any incentives before switching service providers or before you renew your contract for another 2 years.

- Do watch what out for what your favorite celebrities are sporting as their cell phones. Bragging rights are perfectly acceptable while watching Red Carpet coverage before an awards show and you notice a Star who shares the same phone model as you.

Now, how about some of the fashion don'ts for this season...

DON'T
- Don't walk around with your Bluetooth headset in your ear when you aren't on the phone. It looks really geeky, like something out of *Star Trek*.
- Don't wear multiple devices on your hip at the same time. The cell phone super hero utility belt is not as cool as you think. It's a serious chick repellant.
- Don't ask someone, "Can you hear me now?" and expect a serious answer.

RECYCLE YOUR OLD PHONE

This section is in honor of Mr. Al Gore, who inspired me to take action after watching his film "An Inconvenient Truth."

The average person upgrades their cell phone every 16-18 months, corresponding closely to the end of a two-year contract. This makes sense because the end of the contract marks a time when you're free to get a new phone, and since technology changes so rapidly. After years of this repeated pattern, you'll suddenly find yourself with three to four older phones that were at one time the coolest and are now considered dinosaurs. What do you think most people do with those phones?

According to an ecological report from 2005, nearly 130 million cell phones have been discarded in local landfills throughout the United States. As a result, our landfills now contain an excess amount of toxic materials that are polluting the land, water and air just from the discarded cell-phone batteries. Instead of throwing away your unwanted phones, batteries or chargers, please think about recycling, so this material can be disposed of in an environmentally friendly way. Recycling will clean out your desk drawer and allow you to do your part to be kind to the planet at the same time!

There's a variety of locations and drop-off receptacles that are available at places such as wireless retail stores, coffee shops, etc. where you can donate your old phone.

To dispose of your cell phone battery, check the website www.Call2Recycle.org for a listing of drop-off locations. They have noted over 30,000 locations around the country. I found 12 locations within a 15 mile radius of my family's ranch in Montana, so chances are pretty good that there's a location near you!

Recycling your old phones is the responsible thing to do.

Thanks for taking this journey with me as we explore the incredible World of Wireless and the logic behind selecting the right cell phone for you. Hopefully you came away with a sense of understanding – ready to conquer the task at hand with your newly found Decoder Ring.

Use your newfound knowledge to help others as they search for the same information. When someone asks the question "what cell phone should I get?" you will think fondly of this experience and tell them you've got it all figured out!

Part 6

Resources

WHO TO KNOW AND WHERE TO GO FOR MORE INFO

Jen O'Connell – Wireless Expert
Company: www.VoiceOfWireless.com
Book: www.CellPhoneDecoderRing.com
Podcast: www.CandyStorePodcast.com

Battery Drop-off Locations for Recycling
www.Call2Recycle.org

Nationwide Wireless Carrier Websites
Cingular/AT&T: www.cingular.com; www.att.com
Verizon Wireless: www.verizonwireless.com
T-Mobile: www.t-mobile.com
Sprint/Nextel: www.sprint.com

Phone Manufacturer Websites
Motorola: www.hellomoto.com
Nokia: www.nokia.com
Samsung: www.samsung.com/products/wirelessphones/index.htm
LG: www.lgmobilephones.com
Kyocera: www.kyocera-wireless.com
UTStarcom: www.utstar.com/pcd/Support.aspx
palmOne: www.palm.com/us

RIM: www.rim.com ; www.blackberry.com
Apple: www.apple.com/iphone

Sites for Product Reviews
Engadget: www.engadget.com
Phone Scoop: www.phonescoop.com
CNET: www.cnet.com

Trade Publications for Industry News
RCR: www.rcrnews.com
Wireless Week: www.wirelessweek.com

Trade Shows
CES: www.cesweb.org
CTIA: www.ctiawireless.com
CTIA-IT: www.wirelessit.com
3GSM: www.3gsmworldcongress.com

Government Organizations
CTIA: www.ctia.org
FCC: http://wireless.fcc.gov

Glossary

There are over 170 terms documented in this glossary that will help you as you read this book.

TERM	DEFINITION
+1	When using your phone internationally, dial "+1" before you number to call back to the United States.
1G	1st generation – Referring to Analog wireless service.
1XRTT	The acronym that represents 2G technology for CDMA digital technology.
2G	2nd generation – Referring to the digital technologies most commonly considered GSM and CDMA.
2.5G	Half step between 2G and 3G for the GSM technology.
3G	The evolution of CDMA and GSM technologies once the network is able to support multimedia applications at data throughput speeds in excess of 155 kbps.
3 Screens	The term given to represent content available on TV, laptops and cell phones.
4G	The next phase in wireless network evolution, which most likely will represent WiFi, taking wireless multimedia to the next level.
4th Screen	The term given to content that is viewed on a mobile device. 1st screen is the cinema, 2nd screen is TV, 3rd screen is your laptop, and the 4th screen is your mobile device.
A2DP	Bluetooth profile that enables you to listen to stereo music using a Bluetooth-enabled stereo headset.
AAC	A music file format that is commonly used for iPods.

aGPS	Assisted Global Positioning Service – The ability to find a phone's location through the use of the wireless network when a satellite is not available.
America's Tri Band	Access to the wireless network in North America on the 850 MHz and 1900 MHz bands, as well as allowing for the 1800 MHz band for international use.
Application	The generic term given to describe software that can be used for a specific function, such as email.
Band	The term given to describe a group of frequencies that a device uses to communicate, for example range of frequencies that are within close range to the 850 MHz.
Bar phone	Description of a device form factor. A phone that is long and narrow and does not change its shape. Also known as a "candybar" phone.
Battery Meter	The icon on your phone which indicates how much life your battery currently contains before the need to recharge.
Blackberry	The name of an operating system that enables real time access to corporate and personal email. Also, the name given to devices manufactured by RIM.
Blog	Internet page that enables the personal postings of text, video and photos used for social networking.
Bluetooth	Technology that enables two devices to communicate with each other without the need for wires. Most commonly used between headsets and phones.
Bluetooth Capable	A specification that indicates a device can support the Bluetooth technology.
BOGO	Buy One Get One – Sales offer that enables a free item with purchase.
Buddy List	Friends that have access to your Instant Message account that can display their online status by showing their availability.
Buffering	The term given to the intermittent play and pause of audio/video when the size or speed of the file is greater than what the device or network can support.

Cell Phone Decoder Ring

Caller ID	Phone numbers that show on your cell phone display when a call comes in. This can be personalized to display given names and photos for an individual caller.
Candybar	Description of a device form factor. A phone that is long and narrow and does not change its shape. Also known as a "bar phone."
Carrier	A company that offers the ability to communicate over the wireless network. Also known as a "service provider."
CDMA	Code Division Multiple Access – Digital technology used by Verizon Wireless, Sprint and others. Does not use a SIM chip and represents one side of the network ladder.
Cellular	Term given to an industry that relies on network towers to provide voice and data service. Also used interchangeable with the term "wireless."
Centrino	The name given to indicate built-in WiFi access on a laptop.
Character	Number, letter, or symbols used while typing a message or when customizing Caller ID, ect.
CLA	Cigarette Lighter Adaptor – An accessory that charges a phone battery while in a vehicle. Also known as a "Vehicle Power Adaptor."
Clam shell	Description of a device form factor. This phone folds in half to protect the keypad. Also known as a "flip phone."
Cloning	The practice over analog networks of copying a wireless phone number for the purpose of identity theft.
Control Channel	Part of the wireless network that monitors the authentication and validation of service, as well as the part of the network that delivers text messages.
Coverage	Having access to a wireless signal, where your phone is capable of sending and receiving calls and data.
Coverage Area	A geographic landmass that is capable of obtaining a wireless signal.
Converged	When two or more features previously found on separate devices are now contained in one device.

Cross Talk	A conversation you hear that's not your own while you're on an analog network.
Data	Information other than voice, such as text or graphics that is sent over a network.
Data Cable	A cable that connects your device and laptop together for the purpose of transferring files, music, etc.
Data Device	A device that acts like a mini-computer. Most data devices often include a QWERTY keypad and are compatible with laptop applications such as calendar and email. Otherwise known as a "PDA."
Data Speed	The speed at which information travels over the internet, measured in kilobytes (kbps). The faster speeds are found on 3G. Slower data speeds are found on 2G.
Device	Another name for a cell phone that enables a user to send and receive voice and data.
Digital	There are two dominant digital technologies used in North America: GSM and CDMA that offers secure voice and data communications over the wireless network. It's the next generation after Analog. A third digital technology is iDEN, used by Nextel.
Download	The activity of searching for information, games, music or pictures from the internet and bringing it back to a device, usually for an additional fee.
Drivers	Translating software that enables a laptop to talk to another device, such as a PC card.
DRM	Digital Rights Management – The practice of protecting files, as started by the Recording Industry, to ensure proper royalty fees are paid to the artist or owner.
Dual Band	Access to the wireless network in North America on the 850 MHz and 1900 MHz bands.
Dual Flip	Description of a device form factor. This is a device that looks like a normal flip phone, but you can open it a second way to take advantage of a different layout of the keypad – now in QWERTY mode. Also known as "Dual Hinge."

Dual Hinge	Description of a device form factor. This is a device that looks like a normal flip phone, but you can open it a second way to take advantage of a different layout of the keypad – now in QWERTY mode. Also known as "Dual Flip."
Early Adopter	A person who is among the first to use new technology.
Eavesdropping	The practice of an unauthorized person listening to a conversation.
EDGE	2.5G network technology on the GSM side of the network ladder.
Email	Electronic mail sent over the internet.
Email Address	The unique location where electronic mail is sent in order to be received by a specific person.
EV-DO	3G network technology for the CDMA side of the network ladder.
Fast Busy	When phone calls cannot be made or received due to high network congestion.
FCC	Federal Communications Commission – The governing body in the United States that regulates communication industries such as Radio, TV and Wireless.
Features	Items within your device, such as a camera or Bluetooth, that do not require activation of service or additional fees in order to use.
File Sharing	The practice of transferring data between two devices – likely done using a data cable or over Bluetooth.
Flip Phone	Description of a device form factor. This phone folds in half to protect the keypad. Also known as a "Clamshell."
Frequency	The wave that travels across the sky that carries your voice and data over the wireless network. Measured by one crest per second is a "Hertz."
GHz	This stands for Giga Hertz which defines the frequency in the sky. Similar to Mega Hertz (MHz), moving a decimal place over one digit converts 1900 MHz into 1.9 GHz.

GPRS	2G network technology on the GSM side of the ladder.
GPS	Global Positioning Service. A service that can detect your physical location. This service requires direct line of sight to the satellites in the sky in order to be most effective.
GSM	Global Systems for Mobile Communications – Digital technology used by Cingular/AT&T, T-Mobile and others. Uses a SIM chip and represents one side of the network ladder. Also represents the technology standard used predominately across the globe.
Hands Free	The practice of keeping both hands on a vehicle steering wheel while also talking on the phone.
High Speed Data Network	A generic term given to describe the latest in the network evolution for the fastest data throughput available at the time.
Hot Spot	An area where public WiFi service is available.
Hot Swappable	The ability to remove an external memory card without the need to remove the phone battery or turn the phone off.
HSDPA	3G digital technology on the GSM of the ladder that compresses data files so they can travel more efficiently over the network. Works together with the UMTS technology.
iDEN	The digital technology that is used by Nextel over two-way radio networks in true walkie-talkie style. Considered by some as the 3rd digital technology standard supported in the US.
IM	Instant Messaging – A way to communicate real time over the internet, without an archive of the written word. Friends provide their availability to chat from within a Buddy List.
Image Stabilization	The name of a feature within a digital camera that ensures your image will be clear while a picture is taken. Eliminates image blur when a camera may be in slight motion.
IMAP4	The ability to have access to file folders from within a POP3 email account.

IrDA		Inferred – A technology that is a predecessor to Bluetooth that requires direct line of sight to transfer data between 2 machines at close range.
ISP		Internet Service Provider – The name given to a company that offers internet access, email access and/or instant messaging on your laptop – with access to email and IM on a cell phone.
IT		Information Technology – A department that specializes in making technology decisions within a corporation.
Jackknife		Description of a device form factor. This phone is similar to a candybar design, but it rotates to the side on a hinge to make a larger candybar device. Also known as a "swivel."
Kb		Kilobit – A term used to describe the size of data. Smaller than a Mega byte.
Kbps		Kilobits per second – The speed at which data can be sent over the internet. The higher the number, the faster that same amount of data can travel.
Line of Sight		The term given when two devices are required to look at each other in order to communicate – required for IrDA and GPS.
MB		Mega byte – the description of the size of data. Larger than kilobyte.
Mbps		Mega bits per second – the term to describe how fast a file that is 1 MB in size can travel over the internet.
Mega Pixel		A higher quality digital camera found also in cell phones, which has a larger pixel count than a VGA camera. The higher the number of mega pixels, the better the quality of the image when printed.
Memory Card		Removable storage device that saves photos, games, music or anything you would like to save when the memory on your phone is not enough. Also known as "External Memory Card."
Memory Stick		A type of external memory card that gets its name from its particular shape and compatibility. Manufactured by Sony, that supports different form factors called Memory Stick Micro and Duo.
MHz		Mega Hertz. A unit of frequency that makes up spectrum in the sky. Example: 1900 MHz.

Micro SD	A type of external memory card that gets its name from its particular shape and compatibility. Manufactured by Scan Disk – small in size, requires an adaptor for laptop/printer compatibility. Also known as "Transflash."
Mini SD	A type of external memory card that gets its name from its particular shape and compatibility. Manufactured by Scan Disk – small in size, requires an adaptor for laptop/printer compatibility.
Minutes	The amount of time that a person talks on their wireless device, and a point of consideration when selecting a voice rate plan.
MMC	A type of external memory card that gets its name from its particular shape and compatibility.
Mobile to Mobile (M2M)	When one wireless device calls another wireless device. Check with your service provider for rules on M2M calling for pricing discounts.
Modem	The name of a device that is used to enable wireless access to a laptop. A modem can either be fixed or removable from a laptop. This is different from WiFi access.
Module	The name of a device that is used to enable wireless access that is built into a laptop. Most likely it will not support voice – it will be data only. A module is fixed within a laptop. This is different from WiFi access.
MP3	Audio file format commonly used for music.
Multimedia Messaging (MMS)	A text message that contains a picture, audio or video clip. Also known as picture messaging.
MVNO	Mobile Virtual Network Operator – A service provider that does not own the network but rather acts as a reseller.
Network	The commodity that a service provider offers which enables a device to communicate via voice and data, through a series of radio towers.
Next Generation	A generic term that describes the latest advancement of network technology which increases the speed of data access. It's a step up on the network ladder as the network evolves over time.
Optimized	A device that's compatible and approved to work with a particular network or service.

OS	Operating System – The software that a device uses to access features & services. Typically they are carrier-customized and determine ease of use.
OTA	Over the Air – The transfer of information over the internet or through the wireless network.
Pairing	Two devices that have agreed to work only with each other such as a device and Bluetooth headset.
Palm	The name of an operating system for a device that enables access to your calendar and email. Also, the name of a handset manufacturer.
PC Card	Description of a device form factor. The abbreviated name given to a credit card-sized device that enables a laptop to have access to the wireless network. Also known as "PCMCIA."
PCMCIA	Personal Computer Memory Card International Association. The name given to a credit card-sized device that wirelessly enables a laptop to access the cellular network. Also known as a "PC Card."
PCS	Personal Communications Service – A brand name given to the 1900 MHz band width.
PDA	Personal Data Assistant. The name of a device that acts like a mini laptop enabling access to your calendar, emails and other applications.
PDA with a full QWERTY keypad	Description of a device form factor. This is a PDA that uses a full keyboard. It's wider in size than a candybar.
PDA with SureType keypad	Description of a device form factor. This keypad uses one key to represent two letters of the alphabet, making the device look more like a candybar phone and less like a PDA.
Phone	A device with the primary function of making and receiving calls as well as accessing data over the wireless network.
PIN	Personal Identification Number – Also known as a security password.
Pixel	A collection of dots arranged to create an image such as a photo. The higher the pixel count, the better quality the image.

Pocket PC	The name of a Microsoft Operating System that enables full access to Outlook and other MSN applications on your device. These devices support a touch screen.
Pooling	The grouping of minutes, within the same rate plan, for two or more people to access.
POP3	Post Office Protocol 3 – A type of personal email account accessible from a laptop and also available on an alternate device such as a cell phone.
Portal	The name given to a collection of websites that can be accessible from a cell phone and often personalized to an individual.
Postpaid	Wireless service for voice and data access that requires a person to pay after the fact on a reoccurring monthly basis.
Prepaid	Wireless service for voice and data access that requires a person to pay in advance, before talking.
Push to Talk (PTT)	The name given to a service that enables two or more people to communicate with each other in a walkie-talkie style.
Quad Band	Access to the wireless network in North America on the 850 MHz and 1900 MHz bands, as well as allowing for the 900 MHz and 1800 MHz bands for international use.
QWERTY	A keypad that is used on laptops and data devices for ease in typing information quickly.
Reseller	A business model where a company buys service from one provider and offers that service under a different name.
Right sizing	The activity of comparing your actual usage of voice and data against your monthly rate plan. If there is a significant gap either over or under in voice and data usage, then you may wish to consider a new rate plan that matches up closer to your actual usage.
Roaming	When a device no longer is able to communicate with it's original service provider and has to look for an alternate provider to enable the device to make or receive calls.

SD Card	A type of external memory card that gets its name from its particular shape and compatibility. Manufactured by Scan Disk — larger in size.
Service	The ability to make and receive calls and access data over the wireless network.
Service Provider	A company that offers the ability to make and receive calls and access data over the wireless network. Also known as a "carrier."
Sideload	The activity of transferring information (such as music files) between a laptop and a device via a data cable or Bluetooth.
Signal Strength	A meter, found on the screen of your device, which indicates the quality of your wireless signal.
SIM (SIM chip)	Subscriber Identity Module — The brain of a GSM or iDEN phone that contains your phone number, voicemails and messages.
SIM Lock	The practice of restricting a GSM device for activation with a specific GSM carrier.
Single band	Access to the wireless network in North America on the 850 MHz band. Most likely this is an analog device.
Slider	Description of a device form factor. This phone is like a candybar that rises and collapses on itself.
Smartphone	This term has multiple meanings. (1) A generic term given to a PDA with advanced functionality, including calendar and email access. (2) The name of a Microsoft Operating System that enables limited access to Outlook and other MSN applications on your device. (3) A device with QWERTY keypad for messaging.
SMS	An activity that allows a person to send/receive a short note from one mobile device to another. Also known as "Text Messaging."
Spectrum	The grouping of frequencies in the sky to enable the wireless network.
Standby	A device that is ready to make or receive a call.

Standby Time	The maximum amount of hours/days a device can remain powered on (without using) before the battery dies.
Step down	The process a device takes when it tries to access a specific level of the network. When that level is not available, it will then search to find a lower level of the network for continuation of data access.
Streaming Radio/Music	The activity of listening to music on your mobile device, either in the form of stored music files or accessing radio stations over the internet.
Streaming Video/TV	The activity of watching video/TV on your mobile device, either in the form of short clips or through access to live television over the internet.
Stylus	An accessory that resembles a pen and is used on a touch screen device to request a specific command.
SureType	The name of a keypad described as a "modified QWERTY" where one key represents two letters of the alphabet so the device can resemble more of a candybar form factor rather than a PDA.
Swivel	Description of a device form factor. This phone is similar to a candybar design, but it rotates to the side on a hinge to make a larger candybar device. Also known as a "Jackknife."
Symbian	The name of an open standard operating system used on high end data devices.
Sync	A connection between a laptop and a cell phone that enables the transfer of data from one device to another.
Talk Time	The amount of minutes/hours that a phone can be on a call before the battery dies.
TDMA	Time Division Multiple Access. A type of digital technology that is no longer actively supported in the United States and succeeded by the GSM technology. This is not represented on our network ladder.
Tech/Specs	The "technical specifications" that a device supports, which includes digital technology, frequency, applications, features, battery life, etc.
Text Messaging	An activity that allows a person to send/receive a short note from one mobile device to another. Also known as "SMS" – Short Message Service.

Tier 1	A grouping of the nationwide wireless. Includes Cingular/AT&T, Verizon Wireless, Sprint/Nextel and T-Mobile.
Touch Screen	The display on a phone that responds to a command based on the touch of a finger or stylus. Commonly found in PDAs.
Transflash Card	A type of external memory card that gets its name from its particular shape and compatibility. Manufactured by Scan Disk – small in size, requires an adaptor for laptop/printer compatibility. Also known as "MicroSD" and "T-Flash."
Tri Band	Access to the wireless network in North America on the 1900 MHz band, as well as allowing for the 900 MHz and 1800 MHz bands for international use. Often referred to as International Tri Band.
Triple Tap	Using a traditional 12 key phone keypad to send a message or access numbers and letters. This name is derived by the practice of taping on a keypad multiple times to obtain a number, letter or symbol.
Truncate	The function of taking a large text message and breaking it into multiple parts in order to be received on another device.
Type II PC Card Slot	The space on the side of a laptop that supports a removable PC card in order to have access to the wireless network.
UI	User Interface – Software that enables a person to interact with the software on a device.
UMTS	3G network technology on the GSM side of the ladder.
URL	Another name for the address you use when accessing a website.
VGA	A type of low end digital camera found on wireless devices.
Voice of Wireless	Company founded by Jen O'Connell, which is dedicated to educating the public about wireless technology using everyday language.
VPA	Vehicle Power Adaptor– An accessory that charges a phone battery while in a vehicle. Also known as a "Cigarette Lighter Adaptor."

VPN		Virtual Private Network – A company's internal network which protects corporate information behind a firewall.
WiFi		An alternate wireless technology which does not use cellular towers, but rather nodes to enable faster data throughput.
WiMAX		Considered by some service providers to be their 4G strategy on the network ladder, based on WiFi.
Wireless		Term given to an industry that relies on network towers to provide voice and data service. Also used interchangeable with the term "cellular." This term also covers WiFi access.
Wireless 101		A course taught by Voice of Wireless that educates Fortune 500 corporations on the basics of the wireless industry.
Wireless Solution		The combination of the network, services and a device that are grouped together for a person, based on their specific needs.
WMDRM		Windows Media Digital Rights Management – A standard used to define accessibility for downloadable music.

So there you have it… knowledge is power. See you at the next step!

About the Author

Recognized by Wall Street as an expert in Wireless, **Jen O'Connell** has over 12 years experience developing devices, services and networks for nationwide wireless carriers, including Cingular Wireless (AT&T), Verizon Wireless, GTE Wireless and Powertel (T-Mobile). Now she's passing this working knowledge on to you. Author of the *Cell Phone Decoder Ring,* facilitator of "Wireless 101" Training Classes, host of *The Candy Store* Podcast, plus a variety of TV, radio, print and online appearances, Jen has dedicated her professional career to educating the public and Fortune 500 companies about wireless and personal technology. Her extensive knowledge and experience, coupled with her talent for educating while entertaining, have quickly made her the official "Voice of Wireless."

Visit us on the web at www.VoiceOfWireless.com to learn more about Jen O'Connell and the various products and services offered by her company Voice of Wireless, Inc.

Jen O'Connell has a B.A. in Marketing from Winona State University and is a native of Northbrook, IL as well as a frequent visitor to her family's ranch in Montana. Today, she lives in Atlanta, GA with her two cats Bently and Bailey.